The Ethics of Killing Animals

THE ETHICS
OF KILLING ANIMALS

Edited by Tatjana Višak

 and

Robert Garner,

 with

Afterword by Peter Singer

OXFORD

UNIVERSITY PRESS

Oxford University Press is a department of the University of
Oxford. It furthers the University's objective of excellence in research,
scholarship, and education by publishing worldwide.

Oxford   New York
Auckland   Cape Town   Dar es Salaam   Hong Kong   Karachi
Kuala Lumpur   Madrid   Melbourne   Mexico City   Nairobi
New Delhi   Shanghai   Taipei   Toronto

With offices in
Argentina   Austria   Brazil   Chile   Czech Republic   France   Greece
Guatemala   Hungary   Italy   Japan   Poland   Portugal   Singapore
South Korea   Switzerland   Thailand   Turkey   Ukraine   Vietnam

Oxford is a registered trademark of Oxford University Press
in the UK and certain other countries.

Published in the United States of America by
Oxford University Press
198 Madison Avenue, New York, NY 10016

Library of Congress Cataloging-in-Publication Data
The ethics of killing animals / edited by Tatjana Višak and Robert Garner, with afterword by
Peter Singer.
    pages cm
Includes bibliographical references and index.
ISBN 978-0-19-939607-8 (hardcover : alk. paper)—ISBN 978-0-19-939608-5 (pbk. : alk. paper)
1. Animal welfare—Moral and ethical aspects.   2. Animal rights—Moral and ethical aspects.
3. Human-animal relationships—Moral and ethical aspects.   I. Višak, Tatjana, 1974-
II. Garner, Robert, 1960-
HV4708.E846 2016
179′.3—dc23
2015006886

9 8 7 6 5 4 3 2 1
Printed in the United States of America
on acid-free paper

CONTENTS

LIST OF FIGURES

ACKNOWLEDGMENTS

We thank all the contributors, without whom, obviously, this book would not exist. We enjoyed your enthusiasm and our common striving for improvement. We are grateful to Peter Singer for writing the afterword to this volume. Thanks also to the Dike Verlag for their permission to reprint Christine Korsgaard's paper, which is the only one in this volume that was previously published. It appeared before in *Tier und Recht—Entwicklungen und Perspektiven im 21. Jahrhundert* (Animal Law—Developments and Perspectives in the 21st Century), edited by M. Michel, D. Kühne, and J. Hänni (Berlin: Dike Verlag, 2013). Last, but not least, we thank each other and the publisher for our collaboration in what has been a pleasant editing process.

Robert Garner and Tatjana Višak

LIST OF CONTRIBUTORS

Belshaw, Christopher, Philosophy Department, Open University, United Kingdom

Bradley, Ben, Philosophy Department, Syracuse University, United States of America

Cochrane, Alasdair, Department of Politics, University of Sheffield, United Kingdom

Garner, Robert, Department of Politics and International Relations, University of Leicester, United Kingdom

Holtug, Nils, Department of Media, Cognition and Communication, University of Copenhagen, Denmark

Kagan, Shelly, Philosophy Department, Yale University, United States of America

Kaldewaij, Frederike, Philosophy Department, Utrecht University, the Netherlands

Kasperbauer, T. J., Department of Food and Resource Economics, University of Copenhagen, Denmark

Korsgaard, Christine M., Philosophy Department, Harvard University, United States of America

Luper, Steven, Department of Philosophy, Trinity University, United States of America

McMahan, Jeff, Department of Philosophy, Oxford University, United Kingdom

Sandøe, Peter, Department of Food and Resource Economics and Department of Large Animal Sciences, University of Copenhagen, Denmark

Singer, Peter, Department of Philosophy, Princeton University, United States of America and Department of Philosophy, University of Melbourne, Australia

Višak, Tatjana, Philosophy Department, Mannheim University and Philosophy Department, Saarland University, Germany

The Ethics of Killing Animals

Introduction

TATJANA VIŠAK AND ROBERT GARNER

Most moral philosophers would now say that animals have some direct moral worth, and that therefore what we do to them has to be morally justified. It is broadly accepted, and rightly so, that animal welfare matters morally. Many of the animals with which we usually interact, be it as pets or as pests, in our oceans or on our plates, are—or at least were—sentient beings. They are, or were, capable of being bored or engaged, depressed or happy, anxious or at ease. We, human beings, influence these nonhuman animals' welfare in many ways, both directly and indirectly. We caress them and capture them; we buy products made from them and destroy their habitats. Sentient animals deserve our direct moral consideration, because what we do to them matters to them. They can fare well or poorly. They have interests that need to be taken into account. Causing animal suffering is morally problematic. That does not mean that it ought, under all circumstances, to be prevented. Under some conditions it may be morally justified. But it is at least something that requires a moral justification.

Most moral philosophers would also agree that an overwhelming amount of animal suffering that is currently being inflicted is morally unacceptable. Some moral philosophers argue that we ought not to inflict suffering on animals irrespective of the benefits to us, humans, derived from it. Others allow for some exceptions or some weighing of the benefits and harms that a given practice or action brings about. Even though these are theoretically and practically important differences, it is agreed by most moral philosophers—and probably by most members of the public—that the way in which animals are typically raised for human consumption is morally unacceptable.

As Jonathan Safran Foer puts it in his best-selling book *Eating Animals*, "a meat industry that follows the ethics most of us hold . . . cannot deliver the immense amount of cheap meat per capita we currently enjoy" (2009, 229). That means that the current practice of raising and killing animals for food is not in line with prevailing moral ideals about the treatment of animals. For instance, research has revealed that it is common in slaughter facilities for cows to be bled, skinned, and dismembered while conscious (2009, 230). Sometimes this is done deliberately, for practical reasons, such as making sure that the heart is still beating so that the animal bleeds out quicker. On other occasions, it is the result of unskilled workers that are forced to process the animals at a very high speed. In addition, when Temple Grandin first began to quantify the scale of abuse, she reported witnessing "deliberate acts of cruelty occurring on a regular basis" in about one-third of the slaughter facilities (2009, 255). This means that *deliberate* efforts to harm the animals are undertaken by the workers in addition to the suffering that is already inflicted during the regular slaughtering process. This happened in one-third of the facilities during *announced* auditing visits. Foer's vivid description of the everyday horrors of factory farming is based on personal observation and on official reports by governments or the industry. The author and Nobel Prize laureate J. M. Coetzee comments (as quoted on the book's jacket) that "anyone who, after reading Foer's book, continues to consume the industry's products must be without a heart, or impervious to reason, or both."

While such infliction of pain during the killing is widely considered morally wrong, far fewer moral philosophers (and people in general) would want to prohibit the killing of animals if no suffering were involved prior to death. The killing of animals is an important moral issue, not only because of the suffering involved. We kill billions of animals a year for a wide variety of purposes. Globally, the number of land animals killed each year for food has exceeded 65 billion, according to conservative UN Food and Agriculture Organization (FAO) figures. An average American meat-eater is responsible for the suffering and death of about 28 land animals and an estimated 175 aquatic animals per year, totaling over 15,000 individual animals over a 75-year life span (Harish 2012; see also Mohr 2012). Let us assume that the killing is done painlessly and without causing any suffering (even though, as noted above, that is certainly not what happens in most cases when animals are killed) in order to focus on bringing to an end the animal's life *in itself*, and not on any suffering that this may cause for the animal or for others. Is it still morally problematic to end an animal's life? The papers of this book, written by experts in various fields of moral philosophy, contribute toward answering this question.

In the animal ethics literature, the question of the moral value of animal lives has tended to mark a distinction between animal rights and animal welfare positions. The conventional animal rights position, as articulated in this book by Christine Korsgaard and Frederike Kaldewaij, holds that to kill animals, all things being equal, is to infringe their right to life. In fact, there is an alternative rights-based position, as articulated in this book by Alasdair Cochrane, which holds that animal rights and a prohibition on killing animals is not necessarily synonymous. Leaving that debate aside, the major alternative position to an animal rights ethic—the animal welfare ethic—holds that it is morally justifiable to trade off the fundamental interests of animals (including their interest in continued life) in order to secure an aggregative increase in welfare for humans. The central feature of the animal welfare ethic is an insistence that humans are morally superior to animals, but that, since animals have some moral worth, we are not entitled to inflict suffering on them if the human benefit that thereby results is not necessary. The principle of unnecessary suffering, therefore, can be invoked if the level of suffering inflicted on an animal outweighs the benefits likely to be gained by humans. Robert Nozick provides a concise but admirably effective definition of animal welfare when he writes that it constitutes "utilitarianism for animals, Kantianism for people" (1974, 35–42). Sacrificing the interests of animals (including their interest in continued life) for the aggregative welfare, then, is permissible, provided that the benefit is significant enough, but treating humans in the same way is prohibited whatever the benefits that might accrue from so doing.

It is common to draw a distinction in animal ethics between rights-based and utilitarian approaches, both arguing in favor of animal liberation. One can thus distinguish a utilitarian case for animal liberation, as put forward most notably by Peter Singer, from an animal welfare ethic, as described in the previous paragraph. It is true that utilitarian thinkers, such as Jeremy Bentham and John Stuart Mill, were influential in the theorizing of animal welfare. The emphasis placed by utilitarians on the moral value of sentience led to the conclusion that the suffering of animals ought to be taken into account in any valid moral theory. To equate utilitarianism with animal welfare, however, is problematic and potentially misleading. This is because, for utilitarians, the benchmark for moral standing for *both* humans and animals is sentience. That is, while utilitarians (or, to be more precise, act utilitarians) may allow the killing of animals under some circumstances, they also may allow the killing of humans, provided that it maximizes overall welfare in the given circumstances. Following on from this, utilitarians adopt an aggregative principle to determine the rightness or wrongness of an action. Thus, utilitarianism, or at least the version known as act utilitarianism, holds that the rightness or wrongness of an act depends on its effect on overall welfare. The animal welfare ethic, on the other hand, amounts to applying utilitarianism for animals, whereas the ethical treatment of humans is to be judged in an entirely different way, in the sense that significant human interests cannot be traded off in a similar way.

Singer's position is often, incorrectly, equated with animal welfare because he does not use the concept of rights for animals. For Singer, though, to prioritize human interests in the way that the animal welfare ethic does is speciesist, because it is basing moral superiority on species membership rather than on a morally relevant characteristic. That is, present practices have a built-in assumption that nontrivial human interests are not to be counted as part of the utilitarian calculation, but merely defended *whatever the cost to animals*. Because of this, Singer can, and does, arrive at radical conclusions, such as that vegetarianism (barring special cases) is morally obligatory. By contrast, vegetarianism is not essential for animal welfarists. An animal welfarist could not engage in such a cost-benefit analysis where the interests of humans and animals are considered equally.

This book explores the killing of animals from the perspectives of different interrelated fields of ethics, broadly conceived: value theory, normative ethics, and political philosophy. Value theory, quite generally, inquires about what is valuable. For instance, it explores what makes someone's life good or bad for that person. What does someone's quality of life consist in? What makes an individual's death bad for him or her? Can coming into existence be better or worse for an individual than this individual's non-existence? Normative ethics, in turn, is concerned with how we ought to act. From this perspective, one might inquire how animals ought to be treated, and what role welfare considerations should play in agricultural practice. Is it morally permissible to kill an animal? If so, under what conditions?

Some of the papers in this book, in particular in the first part, explore what makes an animal's life good, bad, or neutral for the animal, and what makes an animal's death good, bad, or neutral for it. When we ask the value-theoretical question about what is valuable or good, we are particularly interested in what is *intrinsically* good. The notion can best be explained in contrast to instrumental goods. An instrumental good is good only as a means. One might even say that an instrumental good is not really good, but only a means for getting goods. Money, for instance, is usually considered an instrumental or extrinsic good. It is not good in itself, but it is good for what you can do with it. For instance, you can buy a coat with it. Is the coat good in itself? Arguably not. Rather, it is good for something else that it gives you, such as preventing you from being cold in the winter. Is being prevented from being cold in the winter good in itself? Perhaps it isn't. Perhaps the good thing about staying warm in the winter is that it prevents illnesses. Having a coat thus contributes to your health. Is being healthy a good in itself? Some would argue that it is. Others would argue that it is not. Perhaps even health is only instrumentally good. Perhaps being healthy is only good to the extent that being healthy gives you pleasure and that being unhealthy causes suffering. Is pleasure good in itself? Pleasure might well be good in itself, and suffering might be bad in itself. That does not mean that suffering ought always to be avoided. Visiting the dentist can be instrumentally good for you, because it prevents you from even more suffering at a later time. This does not contradict the claim that suffering as such is intrinsically bad for you. What the intrinsic goods are is a matter of controversy in value theory.

When talking about intrinsic goods, it is common to distinguish between what is *good for* someone and what is good *simpliciter*. A good of the first type is known as a "prudential good." This is what should be captured in an account of welfare. The relationship between prudential goodness and goodness *simpliciter* is controversial. In value theory, it is common to distinguish talk about the individual's "well-being," "welfare," or "prudential good" from talk about what characterizes a good life. Perhaps a good life requires more than only welfare. Perhaps, in order to lead a good life as a normal adult human, one needs to be well off, but also virtuous and concerned about the welfare of others. The value-theoretical contributions to this volume address issues concerning prudential value: What makes an animal's life good or an animal's death bad *for this animal*? Can its existence be better or worse for the animal than never existing?

The questions about the value of life and death for an individual are closely related. According to the most prominent account of the harm of death, the deprivation view, death at a particular time is bad for an individual to the extent that it deprives the individual of what would have made the individual's future life good for her, had she not died at that particular time. In their contributions to this volume, Ben Bradley and Jeff McMahan defend (their favored versions of) the deprivation view in some detail. All other contributors, with the exceptions of Christopher Belshaw and Robert Garner, agree that some version of the deprivation view provides the most plausible account of the harm of death.

In chapter 1, T. J. Kasperbauer and Peter Sandøe provide an analysis of how various theories of animal welfare may be applied to practices of killing animals. First, they sketch the historical development of the concept of animal welfare and present the current debate between prominent theories of animal welfare. Second, they show how animal welfare has been put to use as a criterion for deciding *when* to kill animals for their own sake, and how this has gradually led to debates over *why* animals are killed at all. In common practices of euthanizing sick and elderly pet animals, the animal's welfare is brought forward as a central justification for killing it. In other practices—such as precautionary killing for disease control or the killing of livestock animals that do not serve the purpose of food production, such as the male chicken of laying hens that are routinely killed right after hatching—the killing is not conceived to be in the interest of the animal, or of anyone at all, except perhaps in the economic interest of consumers or producers. In these cases, appeals are made to animal welfare when questioning the acceptability of these killings. Finally, the authors observe an emerging concern about a natural life for animals, which is taken to imply, among other things, that a good animal life is a life of a certain length. In the light of this perfectionist understanding of animal welfare, killing animals can be considered a welfare problem, even if it happens painlessly. This is because it prevents the animal from living out its natural life span.

In chapter 2, Christopher Belshaw draws a distinction between bads that matter morally and bads that don't. He allows that death can be bad for an animal, as indeed it can be bad for a plant, and similarly that it can harm an animal, and thus cause it a drop in well-being or welfare. But it is a further question, he argues, whether death is something we have reason to care about or want to prevent. Belshaw argues that even though death can harm an animal, it is, in most cases, not something that matters morally. In contrast to death, pain not only harms an animal, but it is also something we should care about. The crucial difference, according to Belshaw, is that most animals are not persons, meaning they lack certain capacities, such as rationality and some more demanding form of self-awareness. This means that most animals lack an awareness of their own existence through time and have no desire for continued life. Therefore, according to Belshaw, their continued life does not matter morally. This is different in the case of pain, since pain matters for the animal, and therefore should matter to us.

This leads to two surprising conclusions: first that it isn't bad, in the way that matters, when an animal dies; and second that it is often good, in the way that matters, when an animal dies, because it prevents future pain. To be sure, killing the animal will also often prevent future enjoyment. However, this future enjoyment would not accrue to the same animal, in the relevant sense. Animals, according to Belshaw, are usually not connected in the relevant way to their future selves. Therefore, just as it is unjustified to cause or allow the suffering of some individuals in order to bring about enjoyment for others, so it is unjustified to cause or allow suffering for some time slices of an animal just to bring about enjoyment for other time slices of the same animal. In the case of nonpersons, this analogy between different time slices and different individuals holds, because there is, according to Belshaw, no psychological connectedness between the various time slices. There is some reason, then, according to Belshaw, to kill existing animals, and, further, some reason to prevent new animals from coming to be.

In chapter 3, Ben Bradley focuses on what is allegedly an important distinction in desires. Categorical desires, it has been argued, are necessary for death to be bad. (Belshaw says that they are necessary for death to be bad *in a way that matters*.) Merely conditional desires won't do it. Human beings are able to have categorical desires, while animals allegedly are not. Thus, death is often bad for human beings, but never bad (or bad in a way that matters) for animals. Bradley challenges much of this, querying the distinction on several fronts and suggesting both that animals can have such desires and that such desires aren't necessary for death to be bad. Turning cows into hamburgers, he concludes, is most definitely bad for them. (Bradley assumes, furthermore, that this is bad *in a way that matters*, though he doesn't explicitly add this specification.)

Bradley defends the deprivation view of the harm of death against various criticisms. As explained, the deprivation view entails that ending an animal's life (even if painlessly) harms the animal if and only if the animal would otherwise have had a pleasant future or, more generally, a future with positive welfare. Death may deprive the animal of what it would otherwise have valued, such as pleasant experiences. It is, according to Bradley, irrelevant whether the animal *wants* that future, whether it has particular future-related desires. Therefore, according to Bradley, death can be bad for individuals such as human babies and nonhuman animals even if they lack a conception of their own existence through time and desires for the future. Furthermore, Bradley (as well as Kaldewaij, in this volume) ascribes categorical desires to cows and other animals that we frequently kill, on the basis of capacities that Belshaw considers insufficient or lacking. So there is not only evaluative and normative disagreement about what makes death bad for cows (and whether this matters). There is also factual disagreement about whether the relevant conditions are fulfilled in the case of cows or other animal species.

In chapter 4, Jeff McMahan examines the claim that suffering is worse for animals than death. First, McMahan rejects some possible defenses of the claim that suffering is worse, which appeal to (a) an asymmetry between suffering and enjoyment, (b) the lexical inferiority of animal enjoyments, and (c) the lesser moral status of animals. Then he shows that the view about the harm of death that he defended earlier (i.e. his time-relative interest account) supports the claim that suffering is worse in at least some cases. Consideration of these cases shows the necessity to point out how the particular view about the harm of death deals with possible (or contingent) interests (i.e. interests that do or do not arise depending on what one does). McMahan appeals to the asymmetry between not having a reason to cause individuals with positive welfare to exist and having a reason not to cause individuals with negative welfare to exist. He argues that possible positive welfare lacks reason-giving weight, but it has cancelling weight in the sense that it can counterbalance possible negative welfare. McMahan's time-relative interest account on the harm of death in conjunction with this asymmetry provides only very limited support to the claim that suffering is worse for animals than death.

In chapter 5, Steven Luper takes up the issue of personal identity. In what sense does death deprive an individual of his future? Is the particular cow that one kills now identical, in the relevant sense, to the cow that would have lived a couple of more years if she were not killed? Is psychological connectedness the crucial criterion, as McMahan assumes? According to mentalist accounts only some mental (or psychological) attributes determine what we fundamentally are. If the cow that would have lived in the future would have been psychologically continuous with the cow that we killed, then it is the same cow in the relevant sense. Is it true that we are fundamentally minds or persons, as opposed to animals? If that is true, we are closely connected to an animal (i.e. to our animal body), but we are, at a fundamental level, not animals. Luper, however, brings forward an argument in support of animalism—the view that we are, fundamentally, animals. He points out animalism's implications for the ethics of killing animals.

After this excursion to issues of personal identity, Nils Holtug brings us back to value theory in chapter 6. Holtug's focus is not on the harm of death but on the value of existence. He argues that existence (or rather leading a particular life) can be better or worse for an individual than never existing. Holtug responds to various criticisms of his "value of existence" view. According to these criticisms, existing cannot be better or worse for an individual than never existing, because both states are incommensurable. If I did not exist, I would not have any welfare level or any attributes at all and therefore, according to the critics, non-existence cannot be compared to existence in terms of how it would be *for me*. Holtug rejects that criticism. His view, if correct, can help us account for the intuition that it is bad for an animal to be brought into existence if it has a miserable life. On the other hand, it implies—in line with a famous justification for eating products from "happy animals," the so-called logic of the larder—that granting an animal a brief but good life is better for the animal than having no life at all.

After these value-theoretical explorations, the remaining papers address our moral and political duties towards animals. In Part II, the normative ethical contributions, in contrast to the preceding value-theoretical assessments, focus on telling us not only what is good or bad (for an animal), but also what is morally right or wrong. They aim at telling us what we ought to do. According to many moral theories, value-theoretical information about what benefits or harms an animal is morally relevant. That is, it is relevant for how we ought to act. According to welfarist moral theories, such as utilitarianism, whether our actions are right or wrong is solely determined on the basis of their effects on welfare. According to utilitarianism, an action is right if, and only if, it maximizes welfare. In this volume, Shelly Kagan and Tatjana Višak discuss normative ethical issues about killing animals in relation to utilitarianism. Both Kagan and Višak discuss Peter Singer's position in particular, which is the most influential utilitarian position in animal ethics. Christine Korsgaard and Frederike Kaldewaij, on the other hand, explore our moral duties towards animals from a Kantian perspective, broadly conceived. This makes sense, since Kantianism and utilitarianism are known as the two classical moral theories, which are still very much alive today. Both Korsgaard and Kaldewaij argue—in opposition to Kant and many Kantians—that the most plausible understanding of a Kantian foundation of moral duties implies that we owe direct moral duties to nonhuman animals, including the duty not to kill them.

More specifically, in chapter 7, Višak discusses a prominent argument that justifies the routine killing of animals under certain conditions: Peter Singer's replaceability argument. Roughly, according to the replaceability argument, painlessly killing an animal is morally neutral, if an animal that would not otherwise have existed and whose future life contains at least as much welfare as the future life of the killed animal would have contained, replaces the killed animal. In that way, the welfare loss due to the killing is compensated. Višak argues that utilitarianism need not entail the replaceability argument. Contrary to what has become the standard utilitarian view, this moral theory, according to Višak, can grant animals a stronger protection against killing. Višak's argument is built on the assumption that, contrary to Holtug's position, existing as opposed to never existing, cannot benefit or harm an animal. She proposes a person-affecting, as opposed to impersonal, version of utilitarianism. That means that outcomes should not be evaluated in terms of the quantity of welfare that they contain, but in terms of net aggregate benefit. Višak adds a proposal concerning how to apply person-affecting utilitarianism to cases in which different individuals exist in different outcomes. Her view, if correct, avoids the replaceability argument, as well as some further counter-intuitive implications (known as the "repugnant conclusion" and the "logic of the larder"), while it has more trouble accounting for the intuition that existence, compared to never existing, harms an animal if the animal in question has a miserable life.

In chapter 8, Shelly Kagan takes the assumption that underlies the replaceability argument for granted, if only for the sake of argument. He assumes, with Singer, that outcomes should be evaluated in terms of the quantity of welfare that they contain (i.e. the impersonal view) and that existence, as opposed to never existing, can benefit an individual. However, Kagan argues that individuals that have desires for the future can be excluded from the scope of the replaceability argument. According to Kagan, there is no need for preference utilitarians to complicate their view in various ways in order to exclude (human or nonhuman) persons from replaceability. Kagan argues that even according to a simple version of preference utilitarianism, persons (i.e. individuals with future-directed desires) are not replaceable and should therefore be granted a stronger protection against killing. Like Višak's argument, Kagan's line of thought, if successful, shows that it is possible to grant animals a stronger protection against killing within a utilitarian framework.

In chapter 9, Christine Korsgaard turns to Kantian moral theory and argues that a case for both the moral claims and the legal rights of nonhuman animals can be made on the basis of Kant's own moral and political arguments. According to Korsgaard, Kant's views about the human place in the world—his resistance to the pretensions that human beings have metaphysical knowledge of the way the world is in itself, and the arguments he uses to show that we can construct an objective moral system without such knowledge—require us to acknowledge direct moral duties toward nonhuman animals. She defends both moral and legal rights for animals, including the right to life.

In chapter 10, Frederike Kaldewaij, like Korsgaard, bases duties towards animals on deontological foundations. Kaldewaij also discusses the ethics of killing nonhuman animals from the viewpoint of Kantian constructivism. This label is used for a group of theories that regard the validity of reasons as determined by a procedure that involves the agreement of all concerned. It is traditionally thought that the only beings that have moral status on the basis of such justifications are those that are able to agree with our actions in the relevant sense—those who have the capacity for rational choice or moral autonomy. Kaldewaij argues, however, that if Kantian constructivist arguments are capable of justifying substantive duties to others at all (including to rational beings), they also justify moral duties to nonhuman animals who consciously pursue ends that they want to fulfill. Kaldewaij then discusses the implications of this view with regard to the ethics of killing nonhuman animals. She defends different reasons why killing animals might be thought morally problematic on the basis of such an argument.

The contributions in the field of normative ethics are indeed very much concerned with what actions are morally right or wrong with regard to animals, and with justifying why this is the case. The final two papers, in Part III, seek to politicize normative ethics. For Alasdair Cochrane, rights necessarily demarcate political entitlements, or duties that can be enforced by the state. So rights discourse is inherently political. While Korsgaard and Kaldewaij focus on whether animals ought to possess a right to life through the more traditional Kantian prism, Cochrane addresses the question through an alternative interest-based theory of rights. His starting point is not *whether* animals possess rights, but the much more neglected issue of *which rights* they possess. He develops and defends a two-tiered interest-based approach. In this account, the "abstract" rights of animals are specified by judging which of their basic interests are sufficient to ground duties in others. These abstract rights are then translated into the more specific "concrete" rights of a particular animal in a specific situation by an analysis of all the relevant interests and values at stake in that context. Thus, it may be that many animals, having an interest in continued life, have an abstract and prima facie right to life. This right, however, may not translate into a concrete right to life.

To illustrate this point, Cochrane considers the case of rodents. These are animals that have important interests that impose duties on us, though in concrete terms there may be occasions when there are compelling reasons for overriding their abstract right to life, perhaps on the grounds of self-defense, such as when they invade our homes, for example. Of course, some might argue that this conclusion is arrived at too quickly, not least because the existence of rats does not normally lead to the deaths of human beings. Even accepting this, it is possible to maintain that rodents do not have a concrete right to life, because their interest in continued life is normally weaker than that of humans. This does not mean, Cochrane hastens to add, that the *way* we kill them is of no moral import. Since rodents do have a considerable interest in not suffering, we are obliged to find the most humane form of killing possible.

Robert Garner continues the debate about the moral value of animal lives in the final paper of the book. His approach is inherently political, because he invokes the need to take into account non-ideal constraints on ideal theory. Thus, while it might be the case that a valid ideal theory attaches considerable moral value to animal lives (as many other papers in this book have suggested), Garner argues that such a step is incompatible with the development of a realizable non-ideal theory of justice or morality for animals. This is partly because the killing of animals is such a central part of modern societies that to prohibit much of it, as the traditional animal rights ethic has sought to do, is politically unrealistic and bordering on the utopian.

This does not mean, Garner continues, that non-ideal theory has to embrace an animal welfare ethic that postulates that not only are animal lives of little moral importance, but that even their interests in not suffering can be overridden if significant human benefits are likely to accrue. Rather, Garner argues (following Rawls) that such an ethic is morally impermissible as a non-ideal theory because it fails to ensure that the animals' interest in not suffering at the hands of humans is upheld. This assertion is dependent upon the validity of an interest-based ideal theory which concludes that the interest animals have in avoiding suffering is greater than their interest in continued life. According to Garner, a valid non-ideal theory can therefore be shorn of claims about the moral value of animal lives, but it must ensure that the more fundamental injustice visited on animals—the imposition of suffering—is tackled.

The debate about the issues addressed in this volume continues. All contributors agree that animals deserve our direct moral consideration. All agree that what we do to sentient animals can harm or benefit them. All agree that causing animal suffering is morally problematic and in many cases morally wrong. Most (but not all) agree that killing, even if it is done painlessly, may harm sentient animals, such as mice, fish, cows, or chicken, in a way that matters. Most (but not all) agree that this harm is due to depriving the animal of the future welfare that it would otherwise have experienced. (And if the animal would not have experienced much future welfare anyway, due to our ways of treating it, that is morally problematic and often morally wrong in the first place.) There is some disagreement about whether one should *discount* the future welfare that is being taken away from the animal because of a lack of psychological connectedness between the animal that is killed and its later self. McMahan, Belshaw, and Kaldewaij argue in favor of discounting. Bradley argues against it. The question whether one should discount for a lack of psychological connectedness relates to larger questions about personal identity and what matters in survival. Luper addresses the former issue in this volume. While most contributors agree that whenever we benefit or harm an animal this benefit or harm is morally relevant, they disagree about what specific moral *rights* (if any) animals have, and about what legal animal rights we should strive to realize. Should one's reasonable expectations determine which political reforms to strive for? Which political reforms *can* reasonably be expected in the first place?

With regard to the ethics of killing animals, a plethora of questions are in need of careful consideration. This volume aims at contributing to this debate. As the editors, we would hope that the papers inspire readers to continue to think, and debate, about the killing of animals, to adjust behavior when it is in conflict with philosophical reflection, and to persuade others to act likewise. The animals deserve no less.

PART I

Animals and the Harm of Death

1

Killing as a Welfare Issue

T. J. KASPERBAUER AND PETER SANDØE

1.1. INTRODUCTION

In a recent article, the influential animal welfare scientist Donald Broom claims that "the animal welfare issue is what happens before death, including how [animals] are treated during the last part of their lives, often the pre-slaughter period and then the method by which they are killed" (2011, 126). In contrast, James Yeates asserts that "death is contrary to an animal's interests, i.e., death is a welfare issue" (2010, 239). Broom claims that death is not a welfare issue, because it is only an animal's welfare interests while alive that matter, while Yeates claims that death must be considered a welfare issue, because killing an animal affects the length of the animal's life and thus how much welfare the animal can experience during his or her life.

Broom's claim that death is not a welfare issue reflects what may be viewed as the dominant view of animal welfare, as the term is used in the legal system, in veterinary procedures, and in guidelines for good practices in agricultural production. This paper provides an analysis of this dominant conception of welfare, compares it to prominent alternatives, and explores how these various conceptions can be applied to different practices of killing animals. The aim is to show how one's view on whether killing animals is a welfare issue depends on (1) assumptions concerning the proper use of animals and (2) different conceptions of what counts as a good life for them. In some views of welfare, we argue, killing animals must be considered a welfare issue.

This paper is structured as follows. We start by introducing different views on the nature of animal welfare, focusing particularly on the development of the conception of animal welfare since the publication of the so-called Brambell Report in the United Kingdom in the mid-1960s. We explain how the report's focus on mental states, particularly negative states, has gradually been modified through an expansion of the concept of welfare to include both positive well-being and "natural" lives for animals. The rest of the paper discusses three different kinds of practices where concern for animal welfare is at play when making decisions about whether or not to kill animals. First, we discuss proper euthanasia, where the degree of suffering is used as an indicator for when animals should be killed for their own good. Second, we consider discussions surrounding practices where animals are killed painlessly but not for the sake of the animals themselves and to no serious human end. And third, we discuss the potential role of accounts of welfare that promote a "natural" life for animals, which may treat killing animals more directly as a welfare issue. All three discussions illustrate how questions about killing animals have become intertwined with questions about welfare.

1.2. VIEWS ON WELFARE

"Welfare" is often taken to be synonymous with "well-being" or "quality of life." This can be confusing, however, since these terms denote different things in different contexts. Discussions of animal welfare have traditionally tended to focus on the absence of negative states (pain, frustration, fear, and the like), while discussions of well-being and quality of life, particularly as applied to humans, have tended to focus on the presence of positive states (satisfaction, joy, pleasure, etc.). These trends have shifted recently, however. Discussions of animal welfare increasingly include consideration of positive states, such that today there is little difference in how researchers use the terms "welfare" and "well-being." In this paper, we take these terms to be interchangeable.

The Brambell Report (1965) is often identified as a watershed moment in the discussion of animal welfare. It provided the first clear statement of what animal welfare was, and why farm animals in particular deserved better living conditions. The report was prompted by Ruth Harrison's (1964) critique of intensive animal agriculture, which highlighted the suffering endured by animals reared for human consumption. Whereas livestock had previously only been protected from pointless cruelty, the report initiated protection from overuse and negligence. The recommendations of the report formed the basis of subsequent European animal welfare legislation, and it remains influential today (for more background on the Brambell Report, see Sandøe and Jensen 2011).

A number of the claims found in the report made important contributions to the concept of animal welfare. First, the report claimed that a one-sided focus on efficient animal production was one of the main causes of animal suffering. This claim opposed the widely held belief that highly productive animals necessarily had good welfare. Second, it claimed that intensively reared animals must be allowed to express certain "normal" behaviors in order to satisfy their "behavioral urges" and achieve good welfare. Five such behaviors were identified, labeled the "Brambell freedoms," which included the ability "to stand up, lie down, turn around, groom themselves and stretch their limbs" (Brambell 1965, 13). Third, though the report focused on preventing animal suffering, this was described as involving not just pain, but frustration and discomfort as well. Thus the ability to express "normal" behaviors, as identified in the Brambell freedoms, was seen as preventing frustration, and thereby suffering, which broadened the traditional conception of suffering (Mench 1998). This eventually led to a conception of suffering that is widely accepted today, which defines suffering as any prolonged or intense negative mental state (Broom and Johnson 1993).

Thus the primary conception of welfare that came out of the Brambell Report focused on the status of animals' *mental states*. Specifically, animal welfare was understood as referring to the presence or absence of negative mental states. Popular books connecting animals' mental states to their moral status further supported the focus on mental states. Examples are Peter Singer's *Animal Liberation* (1975) and Donald Griffin's *The Question of Animal Awareness* (1976).

Subsequent scrutiny of the Brambell Report led to recognition of its various limitations, particularly in its prioritization of the prevention of negative mental states. Among mental state accounts of welfare, a distinction can be made between narrow hedonism, where all that matters is absence of suffering, and full hedonism, where both negative states and positive states matter. In essence, the report espoused a narrow version of hedonism, which was applied even to the idea of animals' basic freedoms. For instance, the Brambell freedoms were implicitly understood to be infringed only when their restriction led to suffering.

One of the main problems with narrow hedonism, which we will explore throughout this paper, is that it is obviously inadequate as the *sole* criterion for welfare (Gjerris, Nielsen, and Sandøe 2013, chap. 2). If avoiding suffering was truly all that mattered, then every animal should be killed as soon as possible, since this would ensure the absence of suffering. It is instead the balance of positive and negative states that is usually in question when assessing welfare, even with respect to human beings. Full hedonism is not reflected in the Brambell Report, however, and, as discussed below, it is the narrow Brambell conception that justifies many practices of killing animals, and thus creates problems for these practices when welfare is expanded to include positive states.

In addition to espousing narrow hedonism, the Brambell Report also contained suggestions pointing toward an alternative view, one in which "normal" behaviors had value in their own right. Though the report used the word "normal" to characterize the freedom animals should have to satisfy their behavioral urges, this has subsequently been developed to mean that animals possess "natural" behaviors, specific to each species. Whereas "normal" lends itself to a statistical understanding (what a certain kind of animal typically does), "natural" has metaphysical connotations about the behaviors that characterize certain types of entities.

This idea that animals should be able to express "natural" behaviors ultimately draws from a *perfectionist* view of welfare, according to which good welfare requires the ability to realize certain species-specific potentials. The goal is not to feel well, but to do well. Rather than achieving certain mental states, perfectionist views hold that what is valuable is achieving certain ideals, appropriate to one's species. For instance, another way of reading the Brambell freedoms is that they fostered good animal welfare because they were appropriate for the types of animals in question.

Mental state theories and perfectionist theories both have strong support among ethicists and, as we discuss later, the general public as well. These theories are not exclusive of each other, however, and many ethicists hold hybrid views that also include elements of hedonism. According to Bernie Rollin, for example, identifying an animal's telos (its essential features) is important for animal welfare because "the interests comprising the telos are plausibly what matters most to the animal" (1998, 165). However, Rollin seems to assume a hedonist account of interests:

> Given an animal's *telos*, and the interests that are constitutive thereof, one should not violate those interests. If the animals could be made happier by changing their natures, I see no moral problem in doing so (unless, of course, the changes harm or endanger other animals, humans, or the environment). Telos is not sacred; what is sacred are the interests that follow from it. (1995, 172)

So, what at the end of the day matters for animals, according to Rollin's view, seems to be happiness (i.e. experienced positive mental states), rather than natural living.

Martha Nussbaum's capabilities approach (another perfectionist theory) similarly emphasizes certain mental states in providing a basic level of welfare. According to Nussbaum, "animals are subjects of justice to the extent that individual animals are suffering pain and deprivation" (2006, 357), and although the capabilities approach is not exclusively about suffering, she holds that "the prevention of suffering, both during life and at death, is of crucial importance always" (387). The capabilities approach will be discussed in more detail in section 1.5.

Our discussions in this paper take as their starting point the narrow hedonism of the Brambell conception of welfare, since this is used most frequently when making decisions to kill animals (for more extensive discussion of different views of animal welfare, see Appleby and Sandøe 2002; Sandøe 2011; Sandøe and Christiansen 2007, chap. 3). The next section discusses how welfare questions typically arise when making decisions to kill animals for their own good. We will show how widespread arguments that support killing animals in order to avoid suffering fail to take into account animals' positive well-being, and we will explain what implications this has for the ethics of killing animals.

1.3. CONFLICTS BETWEEN WELFARE AND KILLING

1.3.1. Proper Euthanasia

Most professional guidelines for euthanizing animals define euthanasia as "painless killing." Thus concern for animal welfare is part of the definition. Furthermore, in the case of proper euthanasia, the killing should be done for the sake of the interests of the affected animals. For instance, the American Veterinary Medical Association's (AVMA) guidelines state that euthanasia should be pursued "when death is a welcome event and continued existence is not an attractive option for the animal as perceived by the owner and veterinarian" (AVMA 2013, 7). (Even tough it is common in a veterinary context to use the word "euthanasia" for all painless killing, we restrict the word "euthanasia" here to proper euthanasia, i.e. to cases of (painless) killing that are in the interest of the killed individual.)

The interests of animals are usually defined by reference to mental states, following the narrow Brambell conception of welfare, where what matters is avoiding animal suffering. According to the AVMA, if an animal is experiencing "insurmountable suffering," such that "continuing to live is worse for the animal than death" (2013, 7), then euthanasia is permissible, and in many cases will be obligatory. As the guidelines state, "Prima facie, it is the ethical responsibility of veterinarians to direct animal owners toward euthanasia as a compassionate treatment option when the alternative is prolonged and unrelenting suffering" (9; similar statements can be found in the 2010 EU Directive on the protection of animals used for scientific purposes; see European Commission 2010). It is often considered permissible to kill animals for various reasons (e.g. slaughtering livestock), but when an animal's welfare has become sufficiently poor, proper euthanasia is required.

The difficult, and ethically contentious, aspect of euthanasia is determining when an animal's suffering is so bad that its life *must* be terminated. This is often framed in terms of whether or not it is possible to provide animals a future "life worth living." The point at which a life is not worth living is when it should be terminated. Concluding that an animal's life is *not* worth living is usually due to two factors: (1) an animal is experiencing severe pain or other severe negative states that have no easy remedy, and (2) an animal is afflicted with illness or injury that will ensure future suffering. These criteria appear in the guidelines mentioned above from the AVMA and the EU Directive, the UK's Farm Animal Welfare Council (FAWC 2009), and many other professional organizations. Euthanasia is thus selected when an animal has dropped below a *minimum* threshold for well-being.

However, implicit in the idea of a "life worth living" is a recognition that a painless death cannot be the sole criterion in providing good welfare for animals, even for animals intended for slaughter (like livestock). A commonly used definition of a life worth living is that it has more positive than negative states, considered over the course of a lifetime. According to this definition, an animal in extreme pain might still have a life worth living if the pain is expected to be alleviated and future states to be, on balance, positive. For example, from the point of view of animal welfare, it will be recommended that a young dog with a broken leg should receive treatment instead of being killed, despite short-term suffering. As mentioned above, if avoiding suffering was truly *all* that mattered for animal welfare, then any animal should be slaughtered as soon as possible. Positive well-being matters as well, however, in providing a life worth living.

This illustrates a limitation in the Brambell conception of welfare. Narrow hedonism does not take into account the full range of animals' interests relevant to making decisions about whether or not to euthanize. If there are other goals in providing good animal welfare, like providing positive well-being, then an animal is able to have a net balance of positive states. This would give it a life worth living, and it would remove any obligation to euthanize.

Moreover, if we take positive mental states into account, then we might also call into question cases where a painless death is considered permissible (but not obligatory). The reason for this is that ending a life with net positive well-being might be seen as *a harm*. There is significant debate in the philosophical literature about the harmfulness of death for human beings (Bradley 2009; Bradley, Feldman, and Johansson 2013). The basic idea, as it has been applied to animals, is that death is a harm because it deprives an animal of future benefits (Harman 2011).

A classic objection to the idea that death is a harm is that there is no experiencing subject, after death, that could be the recipient of harm. However, Harman (2011) and others have argued that a benefit does not have to be experienced in order to constitute a benefit. For example, consider a healthy dog in a shelter that is set to be adopted. If the shelter decides instead to kill the dog, the dog is deprived of a significant benefit, even if it is not aware of the loss. Death in such a case may be viewed as a harm, whereas euthanizing the dog in order to prevent prolonged suffering will not.

The idea that death harms an animal has traditionally been kept separate from issues about animal welfare. This is because the death of an animal has been viewed as a post-welfare issue (as indicated in Broom's statement at the beginning of this paper). However, as soon as room is made for the idea that death involves the loss of future positive welfare, then the killing of an animal may be conceptualized as a harm to the animal. The potential harmfulness of death also raises important questions about the moral importance of length of life, which we will return to below. In the next section, we will further explore ways that concern for welfare is employed as a basis for decisions to kill animals, and how this sometimes raises problems for mental state theories of welfare.

1.3.2. The Three Rs in Laboratory Research and the Issue of Killing

Many debates over animal welfare have arisen in the context of laboratory research. Laboratory animals present particularly challenging ethical issues, because they are used to model and understand human diseases and to assess the effects of potentially harmful chemicals, among other things. Painful diseases may therefore be induced in the animals, or they may be exposed to toxic substances. So there is a dilemma in welfare terms between using the animals to help or to protect humans and protecting the animals themselves. This dilemma is typically dealt with by means of the so-called Three Rs: refinement, reduction, and replacement. The authors of these three principles saw them as a way of combining a focus on animal welfare (what they termed "the humanest possible treatment of experimental animals") with scientific progress (Russell and Burch 1959).

The Three Rs (or 3Rs) aim to: (1) refine procedures, so that research can be conducted on live animals with minimal suffering; (2) reduce the number of animals used; and (3) replace animals by using alternative methods that do not involve live animals, as much as possible (for more on the Three Rs, see Olsson et al. 2012). The main goal of the Three Rs is to minimize the total amount of suffering for laboratory animals, while still allowing research goals to be achieved. However, as we shall argue, the issue of killing lurks under the surface.

A widely recognized challenge when implementing the Three Rs is that there can be a trade-off between reduction and refinement. For instance, to ensure statistically valid results where effects of a treatment or the exposure to a toxic substance on animals is compared to a control group, either a large effect or a large number of animals may be required. This means that methods could be designed that require fewer animals to obtain research goals, but that also cause more suffering to each individual animal. Conversely, *more* animals could be used, but with less invasive procedures, thus producing less suffering for individual animals. Since laboratory animals are (with a few exceptions) killed at the end of an experiment, there will in essence be a trade-off between the level of suffering and the number of animals to be killed.

The Three Rs do not directly take potential *positive* states that laboratory animals might experience into account. However, if positive states were included in the moral equation concerning the use of laboratory animals, reduction and replacement would be seen very differently. They would only be required in cases where the net welfare of the animals is negative. Otherwise, neither reduction nor replacement would contribute to increasing the level of animal welfare. The only other reason to pursue reduction and replacement, relating to concern for animals, seems to be to limit the number of animals that eventually have to be killed. So, in essence, the large focus on reduction and replacement seems to reflect a concern about avoiding unnecessary killing of animals, and not just a concern about protecting animal welfare.

In the next section we will explore cases where animals are subject to what is perceived as unnecessary killing.

1.4. KILLING WITHOUT A PROPER PURPOSE

1.4.1. Precautionary Killing

Since 2000, there have been numerous instances where livestock animals have been preemptively culled in order to prevent or control the outbreak of serious diseases, including swine flu, mad cow disease, and many others. Though the animals culled were largely intended for human consumption, and some of these culls were implemented in order to prevent harm to humans, the preemptive actions have been criticized for killing animals unnecessarily. In many cases, other solutions were available but were not selected, primarily for economic reasons.

For instance, in the 2001 outbreak of foot-and-mouth disease (FMD) in the United Kingdom, over four million livestock animals were preemptively culled for disease control purposes (Haydon, Kao, and Kitching 2004). Animals on any farm within three kilometers of an infected area were killed, even if they showed no signs of infection, because different models predicted that this was the best way to control the epidemic. Subsequent analyses, however, have suggested that more effective measures would have included more accurate diagnosis of which animals were infected, more extensive vaccination, and more selective culls. And it seems that the main reason why these measures were not implemented was economic (i.e. the loss of ability to export animal products; see Charleston et al. 2011; Keeling et al. 2003; Smith et al. 2014; Tildesley et al. 2006).

Here, too, we can see the Brambell conception of welfare at work. The only welfare issue with preemptive killing, according to this conception, is whether the animals are given a painless death. And at least part of the reason for choosing preemptive culling seems to be that it provides absolute certainty that the animals will not experience future suffering (e.g. as a result of FMD).

These cases have nonetheless led many to question the practice of killing animals that show no signs of infection. One sort of objection to the FMD cases specifically is that the deaths seemed unnecessary. The animals could have been vaccinated instead of killed. A second, related, objection is that preemptive culling fails to use the animals for their intended purpose: human consumption. Even if one objects to the practice of raising animals for human consumption, one might still find it wasteful to raise an animal for this purpose only to kill it for no reason at all (assuming that this has been true for certain cases of preemptive culling). In sum, if an animal's death has no utility, in either preventing disease or adding to human nutrition, its killing seems unnecessary and wasteful.

These sorts of objections have no place within traditional discussions of welfare that focus on preventing negative mental states. Nor do they fit well within discussions of positive well-being—no case can be made that part of the good life is to be eaten. The idea of proper use may, however, be developed in the light of perfectionist views of welfare. The next section will further explore this idea of proper use.

1.4.2. Useless Animals

Some livestock animals are routinely killed at a young age because they serve no use for human beings. Their existence is unavoidable, but since they are unprofitable for farmers, they are killed as soon as possible. Here we discuss two such animals: bull calves and male laying chicks (for further discussion of these and similar cases, see Franco, Magalhães-Sant'Ana, and Olsson 2014).

Female dairy cattle are, of course, valuable for farmers, as are their female offspring. Their male offspring, however, are only used for breeding and as veal. Few bulls are selected as sires, veal production for many breeds of dairy cattle is not profitable, and since dairy cows must give birth regularly in order to produce milk (about once a year), there is an abundance of bull calves with no usefulness to human beings. As a result, many bull calves are killed at birth.

A similar issue arises with laying hens. Female offspring are valuable for farmers, whereas males are inevitable consequences of reproduction, but unable to provide any human benefits (in principle they could be used for meat production, but they are of a slow-growing breed that cannot compete with birds bred specifically for meat production). Typically, the male offspring of laying hens are killed when less than a day old, shortly after hatching.

These practices arguably adhere to the Brambell conception of welfare more strictly than those discussed above, because they avoid suffering entirely by killing animals at the beginning of life. In doing so, however, these practices also highlight the potential moral importance of *length* of life. These animals are similar to preemptively culled livestock in that they are being prevented from serving the normal human purpose of other members of their species (or other animals of their kind). However, they also are being prevented from living any sort of life at all.

There are two alternative conceptions of welfare we might apply to these cases. First, in a full hedonistic conception of welfare, which includes positive mental states, these practices *deprive* animals of future goods. There is significant debate, on this conception, over whether it is more harmful to kill an animal at an early age than to kill it in its prime (as adult laying hens and dairy cattle are; see Bradley, this volume, and McMahan, this volume). So although a very early death might harm these animals, it is still contested whether this harm is preferable to what would accrue if they lived the typical life of other members of their species.

Second, a perfectionist theory of welfare might conclude that these practices prevent the animals from living a "natural" life. For instance, researchers have argued that the diminishing life spans of dairy cattle (due to changing agricultural practices) prevent them from living a "natural" life, because current trends deviate from historical norms (Bruijnis, Meijboom, and Stassen 2013). And in surveys, ordinary citizens have cited natural-ness (among other concerns) as a reason to change the practice of killing day-old male laying chicks (Leenstra et al. 2011).

The reasoning behind this perfectionist conception seems to be similar to the reasoning described above concerning preemptively culled livestock. Even if a species is reared solely for human consumption, it might be seen as wrong to prevent an individual member of that species from living a life like the other members of its species, especially if this involves killing the animal prematurely. Here the criticism seems particularly biting, because the animals are killed immediately after birth or hatching, and are thus pre-vented from living any life at all. We will return to this issue in section 1.5.

1.4.3. Unwanted Companion Animals

In every case discussed in this section, the animals killed are prevented from living the normal life for members of their species (or other animals of their kind). As just suggested, one reason this might be seen as problematic is that these animals are prevented from living "natural" lives. This same objection can be applied even to companion animals (or pets). Killing a dog prematurely, for instance, might be seen as problematic, because it is not allowed to serve as a companion, like other dogs taken into people's homes.

We can see how this problem arises with "unwanted" companion ani-mals. There are a variety of reasons why a companion animal might become unwanted. Common reasons include the financial burden, allergies, relo-cation, and destructive behavior (DiGiacomo, Arluke, and Patronek 1998; Scarlett et al. 1999). Though exact numbers are not available, many unwanted animals are abandoned or relinquished by their owners, and instead of being adopted or given new homes, they are simply killed. For example, throughout the 1980s and 1990s, an estimated 12–20 million dogs and cats were killed every year in the United States, a rate that has decreased only recently to around 3–4 million (HSUS 2014). Some are euthanized on welfare grounds (e.g. extreme suffering) but a significant number are killed for no reason except that their owners could not keep them, and shelters could not find a home for them.

The welfare problems here are similar to those discussed already. Painlessly killing companions can only be justified within a narrow hedonistic conception of welfare. When we apply alternative theories of welfare, however, this practice appears problematic. No consideration is given to the animals' potential positive well-being, understood either in mentalistic or perfectionist terms.

Killing healthy animals may also be partly responsible for the rise in "no-kill" shelters (usually defined as shelters where 90 percent or more of admitted animals are not killed). In the United States there are reportedly over five hundred cities and towns with at least one no-kill shelter (Saving 90, n.d.), and there's some evidence to indicate that these shelters are to a large degree responsible for the decreased number of dogs and cats being killed in the United States (Frank and Frank 2007). One possible explanation for the rise in no-kill shelters is increased acceptance of the idea that healthy companions should be permitted to serve their purpose as companions. If the sole concern was to reduce suffering, then these animals should just be killed. Shelter life is known to be stressful, and there is no guarantee of finding a good home. This suggests that there is further concern—namely, that the animals should be able to continue serving as companions.

In the next section we will return to perfectionist views, which have more to say about why certain practices of killing animals should be seen as wrong, even when they have no negative impact on well-being defined in terms of mental states.

1.5. LONGEVITY AS A "NATURAL LIFE" QUESTION

For our purposes, the important feature of perfectionist theories is their use of species norms (the common characteristics of members of a species). One relevant feature of species norms, we will suggest, is the normal *length of life* for members of a species. For the purpose of discussing this, we will focus on Martha Nussbaum's capabilities approach.

The core of the capabilities approach, according to Nussbaum, is that "animals are entitled to a wide range of capabilities to function, those that are most essential to a flourishing life, a life worthy of the dignity of each creature" (2006, 392). To determine an animal's capabilities, she says, we must use species norms, because "the species norm (duly evaluated) tells us what the appropriate benchmark is for judging whether a given creature has decent opportunities for flourishing" (365). The aim of the capabilities approach, then, is to identify which of a species' common characteristics are necessary for the flourishing and dignity of any given animal.

The cases discussed in the previous section involved animals who clearly were not allowed to meet their species norm for *length of life*. Nussbaum's position on animal death is that it is harmful insofar as it cuts off activities that are central to an animal's flourishing (385–388). She also explicitly considers length of life as relevant to the capabilities of human beings. The basic capability of human life, she says, means "being able to live to the end of a human life of normal length" (76).

We can apply this "normal length" conception of life to the cases discussed in the previous section. It should be noted first, however, that Nussbaum does not mention species norms when discussing the basic capability of nonhuman life. Rather than using species norms, as she does for humans, she states that the animal capability for life means that "all animals are entitled to continue their lives, whether or not they have such a conscious interest, unless and until pain and decrepitude make death no longer a harm" (2006, 393). Perhaps Nussbaum intends species norms to be implicit in this passage, because pain and decrepitude will vary by species, but it's not clear why she doesn't state this explicitly, as she does for human lives. Despite this oddity, we view our analysis as consistent with the role of species norms in her theory of animal well-being.

Consider again the lives of the male offspring of laying hens and dairy cattle, who are killed as soon as possible after hatching or birth. This is hardly a life of flourishing and dignity. Laying hens and dairy cattle are not allowed to live long lives by human standards, but they are typically allowed to reach adulthood, if not a mature age. One might argue, on perfectionist grounds, that this is a precondition for their flourishing. A certain length of life is appropriate for each species, and it is necessary for good welfare, and this should apply for every member of the species, not just those that provide direct benefits to human beings.

This basic intuition that length of life should be included within a perfectionist conception of welfare also seems to be supported by current "folk conceptions" of welfare. When asked to identify what would be conducive to good animal welfare, ordinary citizens report that good animal welfare requires the expression of "natural" behaviors and living a life appropriate for its species (Lassen, Sandøe, and Forkman 2006; Miele et al. 2011). As mentioned above in the case of day-old male laying chicks, people also seem to consider the length of an animal's life as part of its characteristic and natural functioning (Leenstra et al. 2011).

This is potentially important for a couple of reasons. First, it has been argued that *improving* animal welfare requires significant agreement between researchers and the general public. For instance, the animal welfare scientist Marian Stamp Dawkins claims that "making genuine improvements in animal welfare requires a definition of 'welfare' that everyone can buy into, not a split between a scientific view of welfare and a lay view of welfare" (2008, 942). Most decisions to kill animals, as we have discussed, are made using a conception of welfare that is primarily concerned with avoiding suffering. In order to improve the welfare of animals that are routinely killed, however, perhaps the perfectionist conception of welfare must be given more weight.

Second, the folk perfectionist view might be a threat to the more standard animal welfare perspective that views death as a post-welfare issue. The cases we have discussed indicate that narrow hedonistic theories of welfare might be seen as inadequate for addressing certain practices of killing animals. The folk perfectionist view, by contrast, produces relatively clear conclusions on these cases. Ordinary people might think the "normal" or "natural" life of a laboratory animal, for instance, should be very different from the one allowed by the narrow hedonism that only attempts to prevent suffering. Providing a certain length of life for the animals we have discussed is thus likely to continue to be seen as a welfare issue—and not just a killing issue—by those adhering to a perfectionist theory.

This is only a sketch of the possible implications of perfectionist theories that view length of life as important to welfare. Many other details need to be worked out, of course, including sorting through cases that might present challenges to perfectionist theories. For example, sometimes the living of a "natural" life conflicts with a "natural" end to that life. Zoos, for instance, sometimes allow animals to reproduce naturally, but, as a result, they must cull "surplus" animals before their natural endpoints. Moreover, a "natural" life for animals living in the wild is, on average, very short in duration. If "natural" is taken to mean "mirroring nature," then a natural life for most zoo animals would consist of high mortality rates and likely death at a young age. Determining a natural end to life thus clearly depends on which aspects of nature we include in a perfectionist theory of welfare. Analyzing these sorts of cases will be important for the future development of perfectionist theories.

1.6. CONCLUSION

Killing animals has traditionally not been seen as a welfare issue. However, animal welfare has been viewed as a criterion for when to kill an animal for its own sake. As shown in section 1.3, difficulties in using animal welfare in this way have made the issue of killing animals more ethically salient, and have led to an expansion of the concept of animal welfare.

Furthermore, concern for animal welfare typically comes with assumptions about the proper use of animals, and these assumptions can be challenged by certain killing practices, even when these practices do not directly give rise to suffering or other welfare problems. This, too, requires a rethinking of traditional conceptions of welfare and how they relate to killing animals. As suggested in section 1.4, focusing on a "natural" life for animals might provide a good way of addressing practices that seem to kill animals unnecessarily.

Finally, killing has become a more salient ethical issue with the growing focus (also among the general public) on natural living in relation to animal welfare. In section 1.5 we claimed that one way to think about perfectionist theories of animal welfare, such as Nussbaum's capabilities approach, is that they require a certain length of life for natural living. Insofar as natural living is indeed considered relevant to welfare, we predict that length of life—and thus killing animals—will become even more important to debates over animal welfare. Even if killing animals is not considered an animal welfare issue, it is still deeply intertwined with how welfare is defined and with how limits of acceptable welfare are drawn.

ACKNOWLEDGMENTS

We are grateful to have received helpful comments on this paper from Clare Palmer, Anna Olsson, Tatjana Višak, Nuno Franco, Mickey Gjerris, and Karsten Klint Jensen.

2
Death, Pain, and Animal Life

CHRISTOPHER BELSHAW

Most people believe that many animals can feel pain. And most believe that, other things equal, pain is bad for those who feel it. Most also believe that animals can die. Is death, other things equal, bad for them? A widespread though still controversial view is that a painless death is not bad, at least where nonhuman animals are concerned. I defend a version of that controversial view here. And I advance a further view, also controversial: a painless death is very often good for animals. I argue that it is better for them to die than to live, and indeed that it would be better for them never to exist. If we are thinking of just what is good for these animals, then the fewer of them the better.

These controversial views sit at the center of this paper, occupying, respectively, the second and third sections below. But I begin and end in a somewhat conciliatory manner, exploring ways in which death is bad for animals. In the first section I distinguish between two ways in which such and such might be bad for something. Only one of these, I argue, has implications for morality, or what we ought to do. In the fourth section, I consider how death might be bad for animals in a morally relevant way, but to a lesser degree, typically, than it is bad for people. A shorter final section considers how we should respond to the views advanced here if, as is likely, some doubts remain.

2.1. DEATH IS BAD, I

Rust is bad for tractors. It interferes with their proper functioning, and their being able to do what tractors are supposed to do. It isn't bad just for tractor owners who have fields to plough, and care about their equipment. Similarly, drought is bad for plants, because it threatens their flourishing, independently of the concerns of gardeners and horticulturists. Similarly, again, blindness and arthritis are bad for animals and impact on their well-being. So artifacts, plants, and animals can all be in a better or worse condition, in a good or bad state, depending on what happens to them. And all of them can be damaged, even if only living things can be harmed. We very commonly speak in such ways, and much of what we then say is true.

Should we then oil tractors, water plants, and tend to animals when we come across them? Should we at least do this if we can, if we are not occupied elsewhere, or if other things are equal? Not across the board. Surely it just doesn't matter, unless it matters to tractor owners, or the tractor preservation society, or those wanting a tidy landscape, whether tractors rust or not. Nor is there an important distinction to be drawn here between living and nonliving things. For it similarly doesn't matter whether plants flourish or wither, unless it matters for dietary, aesthetic, or environmental reasons. Or, for those with reservations here, we might instead say it doesn't matter morally, or doesn't matter in a reason-giving or action-guiding way, that tractors rust or that plants wither. We can acknowledge that such things can be in better or worse conditions, can indeed be good or bad exemplars of their kind, and yet maintain that this gives us no reason to care, for their sakes, what condition they are in. And if we have no reason to care about their condition, then we have no reason to act to improve their condition.

It is different with animals. Or at least it is different with those animals that can feel pain. Blindness is not in itself painful, but it can in various ways lead an animal to be in pain. Arthritis is painful in itself, and can cause further pain. Pain matters, or matters morally. There are reasons to care, for the animal's sake, what condition it is in. And so, as almost all of us believe, there can be reasons to act in ways that improve their condition, most obviously and uncontroversially by alleviating and preventing pain.

How strong are these reasons? It's going to vary considerably between cases, but always within limits. First, the reasons can be overridden. No one will have reason to do everything possible to save an animal from pain, just as no one will have reason to do everything possible to save a human being, or even a number of human beings, from pain. Second, though we can always and truly say that someone will have reason, other things equal, to save an animal from pain, we can go further. Since pain matters, we all have reason to prevent or alleviate it, even when other things aren't equal. We all have reason to pay some price, bear some cost, make some sacrifice in order to limit pain. What price is up for debate, but someone who agrees that, other things equal, they should free a hare from a trap, yet insists that if this will cost them their last slice of pizza then the obligation disappears, doesn't take animal pain at all seriously.

Consider now destruction and death. Rust can not only damage a tractor, but can also in time lead to its destruction or disappearance. Drought not only harms plants, but can also kill them.[1] Arthritis and blindness can not only hurt or lead to pain, but can also, at least indirectly, bring about death. But now if it is bad for tractors to be damaged, isn't it bad as well for them to be destroyed? And if it is bad for plants to weaken and to wilt, isn't it also bad for them to die? Similarly, shouldn't we think not only that pain but also that death is bad for animals? In brief, the answer is yes to all three questions, but (and before being less brief) two qualifications can be made. First, we can focus on premature death. Perhaps it isn't bad for a plant to die at the end of its lifespan, but, plausibly, it is bad for it to die ahead of time. And here the same holds for animals. Second, we can elsewhere draw a contrast. Perhaps it isn't bad for a plant to die if it is already in a very bad condition. Still, it is hard to see how death can be good for it. But with animals, where the bad condition can involve pain, it is not infrequently, and not implausibly, suggested that death can sometimes be a good thing, or at least better than life, if it brings misery to an end. So let me suggest that a premature death, when it ends a life that is thus far in a good condition, is bad for the plant or animal concerned.

Does it matter, or matter morally, that things die? Do we have reason, for the plant's or animal's sake, to prevent death, hope for its absence, or regret its occurrence? As it's hard to see why we should care, for their sakes, about the disappearance of tractors, so the same holds for the death of plants. Some healthy ash trees die. This is bad for them, but it isn't at all clear that a bad thing happens, something we should care about, or hope to prevent, if they die. But is it the same with animals? Should we think here, too, that although it's bad for them, it isn't clear that something bad happens, something we should care about, when they die? Or should we think that because animals differ from plants and artifacts where damage and disease is concerned, so too do they differ in death?[2] I'll argue below that there isn't a difference here, and that it isn't, in a way that matters morally, bad for animals to suffer a painless death. That is to come. All I claim now is that we establish nothing of importance, where the serious questions of value and morality are concerned, in arguing merely that death is bad for animals, or that it harms them. Concede that, and we are still at first base. So it doesn't follow, from our establishing such things, that we have any reason not to kill and eat them, or to prevent their death, or to regret their death when it occurs. In many cases, certainly, we do have such reason, for in many cases we are rightly concerned with side effects. But we don't have such reason simply in virtue of allowing that death is (often) bad for those who die.

The point here is not, of course, just that it doesn't follow, from our making such claims about value, or what is a good or bad so and so, or what is good or bad for such and such, that this or that action is morally prohibited or required. What we ought to do, all things considered, has, of course, to take all things into account. And I am ignoring side effects, and thus many things, throughout. The point, rather, is that it doesn't follow, or doesn't always follow, from these claims about value that we have any reason at all to act in ways to preserve or promote that value. In particular, it doesn't follow from the fact of death's being bad for animals that we have any reason at all to save animals from death.

This surely is important. Insist, say where farming or hunting or vivisection are under discussion, that death is bad for animals and you are most likely to be taken as implying that there are therefore reasons at least to control such practices. Add that nothing at all follows, from badness, about morality or what we should do, and your position loses its bite. Curiously, this is just where Ben Bradley ends up (Bradley, this volume). After arguing at length that death is bad for cows, he allows that their deaths might not matter morally, and that it might not be wrong to eat them. But then his insistence that it's bad for cows to be turned into hamburgers is not obviously more potent than the corresponding claim that it's bad for trees to be turned into pencils. It is, in both cases, but our hands remain untied.

2.2. DEATH IS NOT BAD

In one way, death is often bad for animals: it deprives them of experiences they would have valued. But is it bad in the way that matters, or that matters morally?[3] The Epicurean view, notoriously, holds that death is not bad for the one who dies.[4] It focuses on human beings, on death without pain, and on badness that matters. There is, on this view, no reason, for the victim's sake, to prevent or regret a sudden painless death. If this is true for human beings, then it is also, I assume, true for animals. But most people doubt or deny its truth where human beings are concerned. Are they right to have these doubts? And are human beings and nonhuman animals in the same boat here?

Epicureans are hedonists. They hold that pain is bad. But there is no pain at the instant of, or subsequent to, a painless death. So although dying might be bad, death, and being dead, are not. But even granting hedonism, there is a counter: pain is bad, but so too is a loss of pleasure. And when life is and will be good, then death, in taking this goodness away, is bad. This deprivation view is the standard and widely accepted response to the Epicurean challenge.[5] It has merit, but perhaps also some shortcomings. For consider an obvious corollary.

Plausibly, if pain is bad, then more pain is worse. And if the loss of pleasure is bad, then the loss of more pleasure is worse. Thus, on the assumption that life is good, the earlier that death robs you of this life, the worse it is. Most of us are inclined to believe that, other things equal, death at thirty is worse than death at eighty. But is death at three worse than death at thirty? Is death at three weeks worse still? Many doubt this, and think, for example, that if a doctor can save either a mother or her baby, she ought to save the mother. Or if I can pull either a neonate or a teenager from a burning car, then I should opt for the latter. And often the thinking here, roughly, is that the mother and the teenager have more to live for. This is why theirs are the lives to be saved. Though they lose less, they matter more.

The distinction drawn in the previous section helps here. We should agree that, other things equal, the earlier the death, the greater the loss. The baby, assuming she would otherwise grow into adulthood, loses as much good life as the teenager, plus more besides (Bradley, this volume).[6] But is this a loss, or a badness, that matters morally? Is it something that we should want very much to avoid or prevent? Or is it, though bad for the one who dies, not bad in a way that should exercise us? For badness to matter, more is needed. And now the rough thought, that the older victim has more to live for, can be filled in and given shape.

Death, I suggest, is most obviously bad for us when it denies us a future life that, not unreasonably, we want to live. It is most obviously bad when it prevents the completion of worthwhile plans and projects already under way. It is less obviously bad if we are irrationally clinging to life, and also less obviously bad if, though it promises to be in many ways good, life holds no appeal. This might be the case if we are terminally bored, or if we have decided to sacrifice ourselves for some greater cause, or if we are newborns with as yet no grasp of or interest in what lies ahead. There is room for considerable discussion around such cases, and this discussion would be needed, were the aim here to provide a comprehensive account of all the circumstances under which death will be bad, in the way that matters, for the one who dies. But that isn't the aim here. All I need to do is defend the claim that death isn't bad for animals. And all I need for that is to argue that animals fail to meet the necessary conditions of death's being bad, in the way that matters, for the one who dies.

Suppose that animals have a good future life ahead, with a predominance of pleasure over pain. Even so, they don't, I claim, want now to live this life. They have no rounded conception of time, or of themselves as creatures who persist through time, and take no stance regarding their own futures. In short, they are not persons.[7] And even if they can have some rudimentary future-directed desires—a dog wants its master to come home and take it for a walk, —they can't derive from these desires any reasons to go on living. They neither want to live for its own sake, nor do they want things, in the future, for which continuing to live is necessary and which provide reasons for their living on. In brief, they lack what have been called categorical desires. This notion, introduced and developed in a well-known paper by Bernard Williams (1973), has recently been subjected to a number of powerful criticisms (Bradley, this volume[8]). It retains some force nevertheless, and I say a little more about it in the section that follows. Right now, provisionally, and again emphasizing the limits to my concerns, I want to claim that death is bad (in the way that matters) only for persons who have categorical desires.[9] So it isn't bad for animals.

There are then two true claims attendant on the deprivation view. First, death is bad, for the one who dies, in one way, whenever it deprives its victim of a good life. In this sense it can be bad for trees and plants. Second, death is bad, for the one who dies, in the way that matters, only when it deprives its victim of a life that, to the victim, matters. Because living on doesn't matter to animals, because it isn't something they want, it isn't bad, in the relevant sense, for them to die.

2.3. OBJECTIONS

There are several objections that might be made. First, it isn't true that death is bad, in the way that matters, only for those wanting to live. Second, animals do want to live. Third, the picture of reasons and values underpinning these claims is, so far, markedly incomplete.

Elizabeth Harman offers two versions of the first objection. Death, she says, can be bad for a depressed person who, right now, doesn't want to live, and might even want to die. And, in an allegedly more extreme case, it can be bad for someone who "*truly* lives in the moment" (2011, 730).

To address the first case, I need to refine my view slightly. I said that death is most obviously bad for us when it deprives us of a future that we want to live, preventing the completion of worthwhile plans and projects already under way. But we can put weight on the second condition.

Imagine a teenager who was until recently living a good life, and one she wanted to live, with certain medium- to long-term plans and projects in place. Right now she is depressed, with no desire to live. But soon, if she recovers, she will resume her previous good life, taking up again at least some of her previous plans and projects. Death now would be bad for her. (Similarly, death would be bad for us if it occurred during an operation, under anaesthetic, if we assume that we'd resume the same life afterwards.[10]) Harman says that "death does not frustrate her desires and plans" (2011, 730). I say it does, and this is why it is bad, even if the desires and plans are temporarily in abeyance. So, rather than claiming that death is bad only for those who have some categorical desires, we can claim it is bad only when it thwarts the satisfaction of some categorical desires.

The second case is easier. I assume we are considering here someone who lives in the moment long term, and not someone who lives in the moment just for a moment. I assume also that these moments are each of them good. Harman says, "If she dies now, her death is bad for her, even though it frustrates no desires or plans" (730). Shall I simply say the opposite? Not quite. Her death may be bad for her, just as a daffodil's death is bad for the daffodil. But is it bad in a way that matters? Should we make sacrifices to sustain the life of this person who has, and will have, no desire for or interest in her life being sustained? It isn't easy to see why we should do that.[11]

Imagine an extreme version of this example. Scientists build a machine that is able to feel the simple pleasures of the moment. The longer it is switched on, the more pleasure it feels. Switching off or destroying this machine would be bad for it, but it is hard to believe that it could be bad in a way that matters, or that the scientists have reason, for the machine's sake, to sustain its operation.

Consider now those objections that insist there is more going on in animal minds than I allow, and that they do, in the relevant sense, have future plans and desires. These are empirical claims, and as I am in no place to make fine-grained distinctions here, the better strategy is to make concessions where appropriate, dig in where possible, and so limit and mark out the contested ground.

So, although I've spoken very generally of animals, some divisions can be made. And the view here can be seen as composite—I claim death is bad only for creatures having a certain complex psychology, and then claim that animals lack this psychology. But of course I don't want to claim this of human animals. Nor need I claim it of all the others. Perhaps elephants, chimpanzees, some birds and cetaceans have a grasp of the future relevantly similar to ours, such that death is bad for them, and in the way that matters. On the other hand, it is surely beyond any reasonable doubt that most animals—jellyfish, starfish, goldfish, newts, moths, earwigs, spiders, flatworms, silkworms, tapeworms—lack future concern. For these, at least,—indeed for most animals and most kinds of animal—death is certainly not bad.

And now, we can focus down on the disputed area, centering mostly on small and medium-sized mammals and birds of the kinds that often we domesticate, hunt, and eat. Should we think death is bad for these animals? Nothing they do, I suggest, gives grounds for thinking they have anything like the sophisticated psychology attendant on categorical desires, or that they want to live on into the future. In particular, certain superficially complex behaviors (e.g. building nests, storing nuts, migration) fail to offer evidence to the contrary, or to point to these animals having a grasp of the future at all similar to ours.

Consider two situations in more detail. A fox might gnaw off its leg in an endeavor to escape from a trap. Does this show that it wants to live, or even that it puts life above health? I think not. In some sense it ranks a mix of fear and pain above pain alone. There is no reason at all to suggest it either knows or believes it will be killed if it stays put. Some cows will ignore the grass at their feet and walk over a hill to where it is sweeter. Let's assume they want this grass. And suppose this will take them five minutes. Does this show that, before setting off, they have the thought, "We want that grass in the future"? I don't think so. At best, they might think, "we want that grass now" and then they get it as near to now as possible.[12] And I imagine it's the same when a bull walks over a hill to where the cows are sweeter. Contrast their behavior with ours. We might choose to accept or seek out present pain in order to promote future pleasure. There is no reason to think the fox is doing that. And we might decide to set aside pleasure now in order to enjoy more pleasure later. There is no reason to think the cows are doing that.[13]

The third objection will take longer, and provide an opportunity to clarify certain points. It starts with the complaint that my view is obviously extreme and grossly counterintuitive. And that is a mark against it. It is surely clearer that the death of animals can be, and often is, bad, and in a way that matters, than that my argument succeeds.

Yet, to repeat, there are side effects. An animal's death can be bad not only for that animal, but also for us, for plants, or for other animals. And it can be bad for those others in ways that matter. A female badger dies in a cull. Her cubs slowly starve. Thousands are distressed to learn that the last Sumatran tiger has been shot. As there are fewer bees, so fewer plants are pollinated and fewer foodstuffs produced. We might talk here, even if a little loosely, of instrumental value. No one should deny that extending the lives of animals is often of instrumental value—good, and in a way that matters, for others. But my question is one that might be put in terms of personal value, asking whether their living on is good, in the mattering way, just for them.[14] And I've argued that it's not.

Now, though, we might consider a still further question. Is it bad, not for us, for others, or for the animal, but just in itself, or perhaps just for the universe, if a healthy animal should die? This, surely, is a question about intrinsic value. And it needs to be seen as distinct from those focused on so far.[15]

Someone might believe that there can be such value and that a healthy tree, for example, is not only a good tree but a contributor to the good of the universe, making it a better place.[16]And, more generally, lives, or good lives, or pleasure might be thought of as viable bearers of such value. Believe this and you might well, and consistently, believe that it is bad, in a way that matters, that animals die a premature death. For there are then fewer lives, fewer good lives, and less pleasure than there would otherwise be. But this intrinsic value view, in contrast to the view that death is bad for the animals that die, is not so widely held, and would seem to have unwelcome consequences. For if lives, good lives, or pleasure are valuable just in themselves, then it would seem that we have reason not only to preserve the lives in existence, but to add more. Other things equal, the more lives, and the more pleasure, the better. Think here just about human beings: not only infanticide, but also abortion, contraception, and abstinence will have black marks against them. Not so many of those who think it bad when babies die think we should, other things equal, create as many people as possible.

A curious position of someone less supportive of animals than it sometimes might seem can be noted here. Peter Singer has held that, absenting side effects, we do nothing wrong in painlessly killing animals if we ensure they are replaced by similar animals enjoying similar levels of welfare or well-being.[17] But why merely sustain, rather than increase, the number of good lives, or the sums of pleasure? Replacement might find its rationale if we are concerned with instrumental value—my daughter's tortoise dies, so I buy her another—but looks like an uncomfortable half-way house where personal and intrinsic values are concerned.[18]

The aim here is not to establish decisively that any or all of these intrinsic value views are untenable, but it is, first, to suggest they lack intuitive appeal, and second, and more important, to insist that they, and instrumental value views also, need to be distinguished from the view about personal value being explored here. Keep these separate and the claim that the death of animals is not bad, in the way that matters, for those very animals, is perhaps less extreme than at first it might appear.

2.4. DEATH IS GOOD

The second controversial claim is that it is good for animals to die. Or at least this—that it is better for them to die than to live. But why is this controversial? I've already suggested that a familiar and widely accepted idea is that we should in some circumstances put an animal out of its misery, just as we sometimes accept (more readily with animals than with people) that life, now and later, might not be worth living, might be worse than nothing, so that death is preferable. The controversy arises, of course, in suggesting that this is in some way generally true. How generally? Imagine an animal that feels no pain now, and will feel none in the future. Death would not be good for this animal. Death is very plausibly not good for sponges or oysters, just as it isn't good for daffodils, and just as destruction isn't good for tractors. But many animals do feel pain, and will feel pain in the future. For almost all these animals, I want to say, death is good, or at least better than life.[19]

Consider a range of cases involving cats. In the first, a cat is hit by a car. It is in great pain now. If we do nothing it will continue in pain for a week and then die. Assume that we can't give pain relief, but that we can painlessly kill the cat. We should do this.

In the second case, the cat is again hit by a car. Its pain is great. Either it continues in pain for a week and then dies, or we kill it and put it out of its misery now, or we subject it to a series of complex operations that will sustain it in pain for a month, but then lead to a full recovery. We should kill the cat. If death isn't bad for animals then it won't be bad for the cat to die, even when it would otherwise have been restored to good health. But it is bad for the cat to be in pain, and the longer the pain continues the worse it is. So there isn't reason, for the sake of the cat, to subject it to painful operations in order to extend its life, even if that life will thereafter be pleasant.[20] Nor is there reason to let nature take its course. There is a reason, however, to end its pain.

In the third case, the cat is in no pain. But the vet detects a tumor and tells you that severe pain will begin in a month's time. This isn't a reason to kill the cat now. But suppose there won't be an opportunity later. You are planning a long sea voyage with the cat as company, and you know that your squeamishness will prevent you from killing it later, on the boat. So there is reason to kill it, or have it killed, now, before the pain begins.

Domestic animals are in one sense winners here. As we manage their lives, so we can manage their deaths, and in most cases we can choose to kill them only when, and not before, the bad pain begins. But our encounters with wild animals are fewer. I might get only one opportunity to kill a rabbit. If I don't, then, predictably, it will later succumb to disease or fall prey to a cat or fox, or choke or bleed to death because of rubbish we've left behind. The best option, for the rabbit, is that I kill it while I can. And it's the same with foxes, giraffes, bears, lizards, owls, and the like.

I might go further. If death is not bad for them, if they have suffered pains earlier in their lives, if, in practice, a wholly painless death is difficult to bring about, then it is better still if such animals are prevented from coming into existence in the first place. Of course, it may not be all things considered better—there are many good reasons for wanting to keep animals in existence. But if we are thinking just of the individual animals then never existing at all is almost certainly their best option.

I need to distinguish the view here from others with which it might be confused. It isn't a part of the argument, even though it may well too often be true, that animals live, overall, miserable lives. And claims about the goodness of death are not linked to such lives, holding that death, or non-existence, is best when and only when present and future pain outweighs present and future pleasure. Even if there is to be more pleasure than pain, death now might still be best. Nor is it part of the argument that whenever an animal is in any degree of pain, it would, at that time, be better off were it not to exist. Hence my references above to great, or severe, or bad pain. So take moments. An animal might, in a moment, experience only pleasure, or only pain, or some mix of the two. I hold that pleasure can outweigh and compensate for pain in the moment, such that, in spite of the pain, life is at that moment worth living.[21] Then take whole lives. Most will contain some mix of pleasures and pains. Perhaps in too many pains will predominate. But even when this isn't so, there are likely to be moments, or episodes, within the life when pains predominate to such an extent that life, at that moment, is worse than nothing. And if an animal is, in a moment, living a life that is not worth living, then it would be better were it not to exist at that moment, even at the cost of missing out on later moments (and also on earlier moments) when pleasures would predominate. For animals, even though overall pleasure might outweigh pain, it cannot compensate for pain, if the life contains moments that are worse than nothing. Or so I claim.

2.5. AN OBJECTION

Death is better than life, for animals, not only when there is pain now and more pain later, but also when there is pain now but pleasure later, and again when there is pleasure now but pain later. Behind these claims is the view that the lives of animals don't hang together the way human lives do. It can be reasonable for us to endure, or even seek out, pain for the sake of future pleasure. And it can be reasonable for us to continue with our lives, even if we know there will be great pain to come. Not so with animals. For us, but not for them, there is value beyond the moment, with pleasures at one time compensating for pains at another. But there might be a difficulty here.

Think of animal lives as a series of more or less discrete moments, episodes, or stages.[22] Go a little further, and think of them as, in effect, a series of discrete lives. Is it permissible to end a bad life even if this prevents later good lives from coming to be? And is it permissible to end a good life in order to prevent later bad lives from coming to be? At least where human lives are concerned, the second of these, in particular, is surely controversial, with difficulties here even for utilitarian views.[23] There might appear, then, to be the makings of a dilemma. The more integrated the animal life, the greater the reason to hold that pleasures can compensate for pains. The less integrated, the greater the doubt that we can trade off one episode or stage against another.

But human lives are, I've insisted, in general, importantly different. If I kill one person for the sake of others, then typically this is controversial because I kill one who wants to live. It is bad for the one to die, even if this is outweighed by benefits to others. Where animals are concerned, this is not the case. So even if we view the different stages in an animal's life as, in effect, different lives, the objection to killing one for the sake of others lacks its customary force.

2.6. DEATH IS BAD, II

I've allowed that death is bad for animals, just as it is bad for plants, and just as destruction is bad for artifacts, in a way that doesn't matter. I'll now consider further the suggestion made above that for some animals death might also be bad in a way that does matter.

If death is bad for us, insofar as it disrupts our plans and projects and prevents their realization, then it is bad for non-human creatures as well under these conditions. But how bad? It is surely tempting to think that the badness of death, in the way that matters, relates to these plans and projects—their number, their size, the degree of commitment and investment in them, and perhaps also how far they are from completion. Two people, both healthy and both getting pleasure from life, die at thirty rather than eighty. Janet has big plans for the future, and has worked for the past ten years toward those plans. John lives mostly in the moment. Death is worse for her than for him. Suppose an orangutan does look forward to mating again, wants again to be pregnant, recalls her previous infant, and hopes the next one, the whole experience, will be at least as good as the last. This is a rather sophisticated nest of thoughts. But if it can occur, then death now, if it is ever bad for human beings, is to some degree bad for the orangutan.

I might leave it there, except that a not unconnected view, similarly suggesting that animal deaths are of some but limited importance, has attracted considerable criticism. This isn't the place to attempt a detailed exposition and defense of Jeff McMahan's time-relative interest account (TRIA), not least because I am unsure that a full defense can be given, but it is worth some discussion nevertheless.[24]

Imagine someone who lives not "in the moment" but, having limited powers of memory and imagination, in effect starts a new life each day. Call him Creet. It is now Tuesday. Creet has, right now, no interest in what might happen on Thursday. Shall we then say that death on Thursday, say on Thursday afternoon, will not be bad for him? That is certainly a mistake. What we should say is that death will never be very bad for Creet, precisely because his interests stretch over so very little time. But since it will be a little bit bad for him to die on Tuesday afternoon, as that day's projects will remain incomplete, so too will it be a little bit bad for him to die on Thursday afternoon, and for the same reason. The badness of his death is relative to the time, or span of time, over which his interests range.[25]

Elizabeth Harman, I think, agrees with this. She considers Tommy Horse and Billy Cow, both of whom have no interests beyond the next five years, and who thus right now have no interest in being alive beyond these years.[26] Both are ill. If untreated, Billy will very soon begin to suffer, and some months later he will die. If untreated, Tommy will begin to suffer in six years, and some months later he will die. For both of them, treatment will itself be painful, and to be effective it needs to begin now. If treated, they will both enjoy long and happy lives. Harman's view is that for both animals treatment now is justified. Tommy might not care now what will happen in six years' time, but when that time comes, he will, as now, want to live for another five years. So far so good. The problem, for Harman, is squaring this with TRIA. This can't, she claims, explain why treatment is permissible in Tommy's case.

John Broome discusses a similar situation.[27]A baby and a thirty-year-old adult will each die in thirty years if not treated now. TRIA, Broome says, recommends that we treat the adult, as she has, now, by far the greater interest in her future life. But this doesn't seem right. When the disease kicks in, the baby will be thirty, with, say, fifty years ahead, whereas the young adult will be sixty, with only twenty years to go. It seems the baby has more to lose.

Ben Bradley also finds shortcomings with TRIA. Charlie, now three weeks old, will access a trust fund when he is twenty-five. Does it matter whether a thief drains the fund now, or when Charlie is twenty-three? Surely not. But TRIA suggests it does matter, for Charlie now has no interest in his adult life, whereas Charlie later surely does. Again, so much the worse for TRIA.[28]

I don't want to suggest that these writers are attacking a straw man. There may be a version of TRIA out there, and it may be the one advanced by McMahan, that does have such counterintuitive implications. But they are not attacking any view I want to put forward, nor are they attacking every view that relates interests to time. What I need to do here is clarify the view's starting point, and then explore a critical ambiguity.

My claim is that death is bad, in the way that matters, not simply when it deprives us of a good life, but of a good life we want to live. It is bad for animals only if, and insofar as, they want to live. And precisely this is the key, on several readings, to TRIA.[29] So the badness of death is in part a function of the depth and range of the desire for life, such that it is normally less bad for young children than for adults to die, and animal deaths, if bad at all, are bad only to a minor degree. Psychology matters.

But now consider the ambiguity in a key notion here—that of interests.[30] Something can be in my interests (or good for me) even though I have no interest (or wants or desires) regarding it. It would be good for me to stop smoking even though I fully intend to continue. Something can be in a tree's interests—getting watered—even though it has no wants or desires at all. And something can be in a cow's interests—living out its 20- to 25-year lifespan—even though it has no wants or desires beyond the next five years. How are we to understand "interests" in TRIA? I think it has to be in the latter sense. Consider this comment on young human lives: "According to the time-relative interests view, the one month old has a weaker interest in continuing to live than the ten year old has" (Harman 2011, 733). What is very plausible is that the one-month-old has fewer future-directed desires; what is less plausible is that fewer things will be good for it or promote its well-being. But of course it can't be left there. The ambiguity about interests links to a similar ambiguity about badness. Because it has fewer desires, death now is in one sense—the sense that matters—not very bad for the neonate, while in another sense—relating to well-being and life span—it is very bad. We might say here that psychology matters, but biology doesn't. And we can now revisit the alleged problems outlined above.

Assume that Tommy Horse, were he not ill, would have a long and healthy life ahead, and assume also that he has, and will continue to have, some medium-term wants and desires. It will be bad, in both senses, for Tommy to die in six years. Does he want, now, not to die in six years? No. Is it bad for him, now, that he will die in six years? That is perhaps an unfamiliar form of question, but, first, I think the answer is yes, and second, it can be usefully modified and improved. Is it bad for Tommy to be given, now, a slow-acting poison that will cause his death in six years? Yes. And nothing in any defensible version of TRIA suggests otherwise.

Broome's baby and adult have, according to TRIA, different interests in, or desires for, their future lives. TRIA will say that if we can save, right now, only one of them from immediate death, we should save the adult. It doesn't say that if we can save, right now, only one of them from a much later death, we should save the adult. Imagine both are, right now, given a slow-acting poison that will cause death in thirty years, but there is just one shot of the antidote. We should give the antidote to the baby.

Go back to Bradley's trust fund. TRIA makes some claims about psychology. It draws inferences from these claims about the badness of death. But it doesn't generalize, making corresponding inferences about other bads. It is bad for me to be robbed as a baby, since I will later have desires this money will help me satisfy. But if I die as a baby, two things are true. I don't inherit money at twenty-five. But also I don't want money at twenty-five, and I don't have desires this money will help satisfy. Imagine a bridging case. I am left golf clubs in my uncle's will, but they are not to be used until I am twenty-five. Does it matter whether you steal them when I am three weeks or twenty-three? It may matter. Steal them when I am still a baby, and I never develop an interest in golf. But if they are hanging around the house all those years, such an interest may be cultivated. So steal them later and harm is done.

One plausible claim is that interests, in the sense of wants or desires, relate to time. Another is that the death's badness relates to wants and desires. An upshot is that animal death, even if bad, is, in the way that matters, typically less bad than human death. None of the objections to TRIA outlined here give serious challenge to this.

2.7. LIFE, DEATH, AND CAUTION

Perhaps you are tempted to assent to these arguments and agree that death is not bad, in the way that matters, for animals. Perhaps you are tempted to agree also that it's often good for them to die, and so are tempted in turn, at least where side effects have no weight, to begin the killing. But you are not wholly certain the arguments are sound, and so not certain this is what you should do. Should you then hold back, both in thought and action, and, erring on the side of caution, give animals the benefit of the doubt? Stuart Rachels thinks you should. I am unpersuaded.

Rachels believes that animals might have a right to life, and that killing them might be murder. Moreover, he seems to believe also that it might be bad for them to die. 'Might', because there are, he claims, facts on both sides. Animals have experiences and desires, and they can have worthwhile lives. Still, they don't desire to live, and they have no concept of life and death. Neither of these views indisputably outweighs the other. Given that there's something approaching a stand-off here, we should put a brake on all anti-animal activities. Yet, Rachels says, "the Argument from Caution won't persuade many people, because most people think they *know* whether animals have a right to life." However, he continues, "I do not share their certainty. The Argument from Caution, I think, is excellent, even if it changes few minds" (2011, 894).[31]

The argument isn't excellent. Let's allow the points about uncertainty. The objection concerns the focus just on the possibility of error, and in overlooking the size of the costs that might be incurred. Suppose we falsely believe that killing animals is murder, and so keep them alive. Very likely we condemn them to considerable suffering, either now or in the relatively near future. Suppose instead we falsely believe they have no right to life, and so kill them. They lose their lives, but painlessly, and miss nothing, regret nothing. This looks like the lesser evil. But it is objected that it is precisely in their regretting nothing, missing nothing, feeling and thinking nothing, now and forever after that the evil of death resides. So it will be said. But "precisely" here is a mere rhetorical flourish, a gesture towards cogency, making no real contribution to debate. And even if we suppose that death is sometimes bad, in the way that matters, for people and for animals, there is still going to be a puzzle as to just how bad, in relation to pain, it is.

NOTES

1. I mean to imply here that killing a thing is often a form of, rather than an alternative to, harming it. As Joel Feinberg notes, everyday discourse might deny this: 'If a murderer is asked whether he has harmed his victim, he might well reply: 'Harmed him? Hell no; I killed him outright!'" (quoted in Fischer 1993, 171–2).

2. Here's an argument to this effect. If pain matters morally, then those things able to suffer pain matter morally. And things that matter morally have moral status. But if something has moral status, then harming that thing matters morally. As I allow, death harms an animal. So we have some reason, which admittedly can be overridden, to save an animal from death. This argument is defective, however. Suppose we allow that animals have moral status. All that follows is that some harms matter. Their pains matter. It doesn't follow that all harms matter. It doesn't follow that their deaths matter. See Sachs 2011 for reservations about moral status generally, and Harman 2011 (728) for a version of this defective argument.

3. Hereafter context will determine whether the qualifications, when unstated, are implied.

4. See Epicurus 1926 and, for important discussion, Warren 2004.

5. For exposition and discussion see the introduction and several papers in Fischer 1993 and Taylor 2013.

6. For a robust defense of this view, often described now as the "life comparative account," see also Bradley 2009.

7. For an important discussion and a more nuanced view than that offered here, see Singer 1993, 110–20.

8. For trenchant, even if not decisive, criticism, see also Bradley and McDaniel 2013. See also Broome 1999, 234–38. Some of the difficulties are also explored in Belshaw 2013.

9. Given that only persons are able to have categorical desires, reference to persons here is in fact redundant. But it has, I think, an important heuristic function nevertheless.

10. For a fuller discussion of the issues here, see Belshaw 2013, 280–82. And the importance of the same life, in the biographical sense, underlies McMahan's discussion of the thought experiment, *The Cure,* which similarly limits the badness of death. (See McMahan 2002, 77; see also Belshaw 2005, 48–49; Bradley 2009, 117–21.)

11. In support of this, first think about what living in the moment would be like. Someone living in the moment presumably can't follow a murder mystery on the television, can't properly read a novel, and can't join in on any but the most rudimentary conversations. Second, recall that an alleged benefit of choosing to live in the moment, among at least some Buddhists, is to be rid of the fear and badness of death.

12. See Bradley, this volume, for this example: "Wouldn't the cow who sees some grass over on the hill, and walks over to the hill to eat the grass, have a desire to eat grass in the future that explains why it walks over to the hill?"

13. Claims about animals having morally important future concerns are widespread. Thus "horses do have desires, and it is apparent that they often act in order to continue living" (Bradley and McDaniel 2013, 128). Also, "Sentient beings, by virtue of their being sentient, have an interest in remaining alive; that is, they prefer, want or desire to stay alive" (Gary Francione, in Francione and Garner 2010, 15). Mark Rowlands offers a subtle, but still I think unsuccessful argument in which, having located animals and their futures between inanimate objects, on the one hand, and human beings on the other, he infers that their futures matter, though less so than do ours (Rowlands 2002, 76–99).

14. Two clarifications are needed here. First, personal values are not meant as subjective values. Second, personal values are not restricted to persons. See Dworkin

1993 for a very useful discussion of the taxonomy of value, and for influences on the distinctions made here.

15. In arguing that death isn't bad for the animal that dies, Alistair Norcross (2012) opts for a position similar to that advanced here. But, perhaps with too little argument, he claims also that in limiting global well-being, such deaths are morally significant.

16. "If you kill a magnificent tree by pouring poison on its roots, you make the world a worse place" (Glover 2006, 46). This sort of thing is often and easily said, but isn't, I think, as straightforward as it first appears. We are one good tree down, but imagine (a) the tree is lost in a storm and (b) there are no bad side effects. Where now is this badness for the world coming from?

17. See Singer 1993, 121–31. But note, first, that Singer in part anticipates the objection made here, and second, (and as very thoroughly explored in Višak 2013) that he has since shifted his underlying position considerably.

18. See Dworkin 1993, 74, for the claim (not, I think, well supported) that there is a species of intrinsic value that isn't open to this incrementality objection.

19. Someone might opt for the second locution on the grounds that being dead isn't intrinsically good. But then it isn't intrinsically bad either, and yet there is little compunction about holding that death is sometimes bad.

20. Harman disagrees: "Consider a young cat that could lead a long happy life if it is given serious surgery that would give it quite a bit of pain (even with painkillers) for a few days, followed by months of serious discomfort. Otherwise the cat will die within a few days, without experiencing much discomfort. In this case it is permissible to do the surgery" (2011, 732). Unfortunately, she says nothing at all to explain this.

21. My view should be distinguished, then, from that of David Benatar (2006), who appears to hold that any amount of pain is enough to render life not worth living.

22. Harman (2011, 731–33) discusses the relation between stage views and claims about moral status at some length.

23. Not, however, for a thoroughgoing negative utilitarian, who assigns no value to pleasure. Yet that, I take it, is an implausible position.

24. McMahan's account is most fully presented in McMahan 2002, chapters 2 and 3. Bradley (2009, 119–46) discusses it at some length, mentioning there my earlier alleged endorsements. Harman (2011, 733–35) describes it, not as an account but a view, and hence TRIV.

25. To clarify, the emphasis here is on individual interests, and not on their collective spread. Creet lives for a hundred years and always has some interests that range over a week. Croot lives for a hundred years and always has some interest that range over thirty years. Other things equal, death is much worse for Croot.

26. Harman 2011, 734. The claims here seem to involve an implausible account of animal life. I find it impossible to believe that these animals have any notion of, concerns for, or interest in, a future five years ahead. But if they will live for, say, another twenty years, then it is in their interests now that things don't happen now that will hamper them in twenty years. Either way, a five-year horizon seems inappropriate.

27. Broome 2004, 251. In the preceding discussion (249–51), and noting McMahan's indebtedness to Parfit, Broome talks about how much well-being matters to a person. As with interests, this notion of mattering is surely importantly ambiguous.

28. Bradley 2009, with the trust fund example on 136. In a footnote, Bradley describes Broome's as a "more complicated counterexample."

29. Thus, on McMahan's account, "the badness of death is not simply a matter of how good the lost life would have been. Rather, it matters also what the being's *psychological relationship* is with that potential future life" (Harman 2011, 733). And Bradley concedes that there at least seems to be something in this: "Surely there is some kernel of truth in the vicinity of TRIA that has attracted thoughtful philosophers" (2009, 145–46).

30. I've used the term on some occasions earlier. If any of those uses were unclear they should, after this short discussion, be clearer.

31. It might be objected that I misrepresent the argument, as it deals just with vegetarianism, and not with further contexts involving animal killing. But, first, caution about factory farming is vastly more defensible than caution about meat eating generally, and, second, it isn't easy to see how vegetarianism itself can be playing any particular role here.

3

Is Death Bad for a Cow?

BEN BRADLEY

3.1. INTRODUCTION

Whether or not death is bad for a cow seems to be determined by what sort of mental lives cows have. According to an article in the *Sunday Times*, scientists have found that "cows have a secret mental life in which they bear grudges, nurture friendships and become excited over intellectual challenges" (Leake 2005). If cows have friends, bear grudges, and get excited over intellectual challenges, then they have mental lives of some sort. But this does not by itself settle the question of whether death is bad for cows, or whether we may permissibly kill them. Some allege that while cows have some mental states, they do not have the ones that are required in order for death to be bad for them. In what follows I will evaluate these claims. Since our knowledge about cows' mental lives may change over time, and we might come to discover that they are more or less sophisticated than we now think, my arguments will not turn on our current understanding of cows, but rather on more general connections between value and mental capacities. I will sometimes grant my opponents pessimistic and perhaps unwarranted assumptions about cow psychology in order to focus on these general connections.

There is a straightforward argument for the conclusion that death is bad for cows in at least some instances (DeGrazia 1996, 231–32; Carruthers 1992, 81; E. Harman 2011, 728; Norcross 2012, 466). It goes like this:

1. Death is bad for an individual if and only if it makes that individual's lifetime well-being level lower than it would otherwise have been.

2. Death sometimes makes a cow's lifetime well-being level lower than it would have been.
3. Death, therefore, is sometimes bad for a cow.

The first premise is just a statement of a generic deprivation account of the badness of death.[1] The feature of this statement that is crucial for our purposes is the notion of a lifetime level of well-being. Your lifetime well-being level is the value for you of your whole life.

The second premise might be justified in a number of ways. Let us suppose that among the goods of life is pleasure, and that among the bads of life is pain. Almost everyone thinks this is true, even if most of us think that other things are good and bad as well. Cows can experience pleasure and pain. If pleasure is among the goods of life, then, *ceteris paribus*, when a cow gets more pleasure, its lifetime well-being level is higher. Death deprives at least some cows of a future that contains some pleasures and comparatively less pain. For those cows, death makes the cow's lifetime well-being level lower than it would have been if the cow had survived.

Those who argue that death is not bad for cows can be divided into those that reject premise 1 and those that reject premise 2. Those who reject premise 1 claim that it is not enough that death makes its victim's lifetime well-being level lower than it would have been; something else, in addition to (or instead of) deprivation, is required for death to be a misfortune. Those who reject premise 2 claim that cows do not have lifetime well-being levels at all. I will consider these responses in turn. But first let me mention some responses to this argument that I won't be discussing here.

Epicurus and Lucretius are famous for giving arguments for the claim that death is not bad for people. Epicurus (1964) argued that death is not bad for us because when death comes, we no longer exist, so there is no subject of harm. Lucretius (1965) argued that death is not bad for us because a future in which we do not exist is relevantly like a past in which we do not exist, and since it is not bad to be born later than we might have been, it is also not bad to die earlier. Some contemporary philosophers find these arguments compelling (e.g. Rorty 1983). If sound, they might provide reason to reject premise 1. I'm not going to address these arguments here, however. I'm supposing that there is no *general* problem with saying that death is bad. What I'm interested in is the idea that although death is bad for some individuals, such as most adult humans, it's not bad for cows.

Pessimists such as David Benatar argue that no actual person lives a life that is worth living on the whole. Life is full of pain. We get sick, we get hungry and tired, we spend great amounts of our lives in discomfort of some sort. Many of us think that the good times in life make it worthwhile to have been born, but we are wrong (Benatar 2006, chap. 3). I think if Benatar is right, we have good reason to reject premise 2, as long as cows are like humans in the relevant respects. I see no reason to think he is right, but I don't wish to discuss this here, because again, if he is right, it would mean that death is not bad for anybody, whether human or not.

Finally, I will not discuss here the views of those who think that nonhumans do not matter at all, morally speaking. As far as I can tell, none of the philosophers to whom I am responding think that animal suffering does not matter. For example, Ruth Cigman argues that we have no obligation not to kill animals painlessly, but nevertheless "we have an obligation not to inflict gratuitous suffering on animals" (1981, 50). The thought is that there is something special about *death*, in virtue of which, while it may be very bad for a cow to be tortured, it is not bad for it to die.

3.2. THE DESIRE FOR LIFE AND THE BADNESS OF DEATH

Chris Belshaw claims that "a desire for life is necessary if death is to be bad" (2009, 115). He qualifies this claim in an important way, as we will see. But if Belshaw is right, the deprivation account of death's badness is false, and the argument fails. Further, if cows lack a desire to live, then death is not bad for them.

Do cows desire to live? According to the People for the Ethical Treatment of Animals, "cows value their lives and don't want to die" (PETA, n.d.). It is unclear what the basis for this claim is, but PETA also recounts a story about a cow that jumped over a fence while being led to the slaughterhouse, and ran a long way away. It was then given sanctuary by an animal rights group. Does this cow's quest for freedom show that the cow wanted to live? I think this is a reasonable explanation of the cow's behavior. Behavior can only constitute *prima facie* evidence of desires, but as Mary Midgley argues, given what we know about similarities between the biological and social characteristics of humans and other animals, "reasons must be found for *refusing* to say" that animals lack the mental states humans have (1983, 134).[2]

While I am intrigued by the question of whether cows have desires, and which ones they have, I am more interested in the claim that a desire to live is necessary for death to be bad. Belshaw offers no argument for the claim. Of course, desire-based views of value are popular, so the claim might be motivated by some such general view about value. But even if a desire-based view of value were true, death could still be bad for an individual that lacked a desire to live. If that individual had *other* desires that would be frustrated by death, death could be bad for her or him. So why think that this one particular desire is necessary?

Perhaps the answer can be found in a distinction between types of desires. Belshaw says that what is really required for death to be bad is not a desire to live, but a *categorical* desire.[3] Here Belshaw follows Bernard Williams, who introduced the notion of a categorical desire in his famous 1973 paper "The Makropulos Case." A categorical desire is supposed to be one that is not conditional on being alive, and therefore can provide both motivation and justification to continue living.

What is it for a desire to be conditional on being alive? To understand this, we must understand the nature of conditional desire. It seems that many desires have conditions—we want something on the condition that some other thing is the case. Charlton wants to eat the soylent green later, on the condition that he is hungry later; if he is not hungry later, then not having some soylent green doesn't frustrate his desire for soylent green. Or, Charlton wants to eat the soylent green later, on the condition that soylent green is something permissible to eat; even if he eats the soylent green later, his desire won't be fulfilled, if soylent turns out to be impermissible to eat. In these cases, a desire is conditional on something that fails to obtain. When this happens, the desire is neither satisfied nor frustrated; it is cancelled (McDaniel and Bradley 2008). It is, in one way, as if the desire never happened.

If this is right, then a categorical desire might be understood to be a desire such that if the desirer were to die, the desire would *not* thereby be cancelled. And we might think, therefore, that some such desire is necessary in order for death to be bad for its victim—at least this is what Williams and Belshaw conclude. If someone had no categorical desires, then upon death all of that person's desires would be cancelled; thus death would frustrate no desires of the person who dies. The following sort of example is supposed to illustrate how this might happen. Someone is very tired of living, and is confined to her bed, but still desires that someone bring her some painkillers. Why? She wants the painkillers, but only on the condition that she is still alive. The painkillers are not motivating her to stay alive, nor giving her any reason to stay alive. If all her desires were like that, and she were to die, we might say that she is no worse off than she would have been if she had lived and had the painkillers.

So now our questions are: can cows have categorical desires, and are categorical desires really necessary for death to be bad? I see no reason to think that cows cannot have categorical desires, provided they can have desires at all. Suppose a cow sees some grass and forms a desire to eat it; while it is moving towards the grass, it is hit by a cannonball and dies instantly. Was the cow's desire for the grass conditional on being alive, like the desire for painkillers in the case of the person who is tired of living? This is implausible. There is no reason to think that the cow wants the grass on the condition that it is alive. It just wants the grass. Its desire does not seem to be conditional on anything at all. Those who are suspicious of the mental lives of cows should, if anything, be *more* suspicious that they have *conditional* desires than that they have categorical ones.

Are categorical desires necessary for death to be bad? The first thing to note is that this claim requires the truth of a desire-based theory of well-being. I think these theories are false, but right now we're not going to settle which is the correct theory of well-being. So let us suppose for now that some desire-based theory of well-being is true, so that differences in well-being must be explained by appeal to differences in the satisfaction or frustration of desires. We have now seen that there is a third thing that can happen to a desire: it can be cancelled. According to a desire-based theory, it is a good thing to have one's desires satisfied, and a bad thing to have them frustrated. What about having a desire cancelled? The natural thing to say is that it is neither good nor bad when a desire is cancelled. It is as if the desire never happened. Now suppose that death cancels some non-categorical desire that would have been satisfied if the victim had not died. Isn't it then bad for the victim to have died? After all, it deprives the victim of the goodness of having a desire satisfied. Preventing the satisfaction of a desire, by cancelling it, is a bad thing, even if it is not as bad as frustrating the desire.

In the case of the person who has tired of living, it seems that death is not a bad thing, even though death cancels a desire for painkillers that would have been satisfied if she had not died. That is not because her desire for painkillers is not categorical, but rather because her desire is an *extrinsic* desire. She desires painkillers on the condition that getting them is necessary to relieve her pain, which we may suppose she desires intrinsically. Defenders of desire-based theories (e.g. Brandt 1979; Heathwood 2005) typically claim, for just this reason, that it is only intrinsic desires that affect well-being. Getting painkillers and dying are both ways to satisfy the intrinsic desire to avoid pain.

So I suspect that a more plausible thought about desires and death is that an *intrinsic* desire is necessary for death to be bad; intrinsic desires are what the person who is tired of living lacks (beyond the desire to be free from pain, which is equally satisfied by dying). But cows can have intrinsic desires. They intrinsically desire grass—or eating grass, or the sensations they get from eating grass. And furthermore, it is not clear that *actually having* an intrinsic desire is necessary for death to be bad. Suppose that the person who is tired of living would, if she survived just a few more days, develop some intrinsic desires for things, and that they would be satisfied. If she dies before those desires form, isn't that a misfortune for her? After all, if she formed the desires and they were satisfied, this would be a good thing for her, so her death makes things go worse for her. Why should we be so concerned with the desires the victim has at the time of death, and ignore the desires she would have had if she lived?[4]

Another kind of desire that cows allegedly lack is desires about the future. Cows are said to lack such desires because they lack the capacity to conceive of themselves as existing through time. If a being cannot conceive of itself as existing through time, then it cannot conceive of its future, and so cannot have desires about it. If a being can't desire to have a certain sort of future, then it can't have such desires frustrated. So desire satisfaction-ists can argue that death is less bad for cows than for adult people, since the death of an adult person frustrates more desires than does the death of a cow. Here is what Peter Singer says about the distinction between beings that grasp their futures and those that don't:

> Those who understand that they exist over time can have preferences relating to the future, and those that cannot understand that they exist over time cannot. Thus the former have more to lose than the latter, and so, other things being equal, it is worse to end its life. (1999, 310)

Singer's concern is with the wrongness of killing, not the badness of death for its victim. What he says is correct; if we presuppose a desire—or preference-based theory of welfare—and if cows have no desires about their futures, the death of a cow is *less* bad than that of a typical adult human. But why should we think cows lack these desires?[5] Wouldn't the cow who sees some grass over on the hill, and walks over to the hill to eat the grass, have a desire to eat grass in the future that explains why it walks over to the hill? Can a creature that has memories and anticipates things really be said to live purely "in the moment"? (DeGrazia 1996, 169–171). Singer and others might have in mind more distant future-directed desires, or desires that concern one's life as a whole. Cows do not put money into 401(k) plans, build bomb shelters, or visit the dentist, as we do when thinking about our more distant futures. As Belshaw says, "wanting a mate right now isn't the same as, or the same sort of thing as, wanting to settle down and raise a family" (2012, 278). Perhaps humans have *more* future-directed desires than cows do. If so, then if some desire-based axiology is true, the death of a cow may typically be *less bad* than that of a human. Nevertheless, the death of a cow may still be very bad for it, since, in addition to frustrating the perhaps limited number of future-directed desires it has when it dies, it prevents the cow from forming and satisfying preferences in the future, and such satisfactions would be good for the cow. So we need further argument to establish that a cow's death is not bad for it.

The story I have told about categorical desires makes it fairly easy to have a categorical desire. Some seem to think it is much more difficult to have such desires. Here is what Ruth Cigman says:

> The subject of a categorical desire must either understand death as a condition which closes a possible future forever, and leaves behind one a world in which one has no part as an agent or conscious being of any sort; or he must grasp, and then reject, this conception of death, in favor of a belief in immortality. Either way, the radical and exclusive nature of the transition from life to death must be understood. (1981, 58–59)

It is not clear why Cigman thinks having a categorical desire requires the desirer to understand death in this way. But arguments about what a categorical desire is are not very interesting, because "categorical desire" is just a technical term. We can consider her view on its own merits, whether or not we think of it as a view about categorical desire. Must one be capable of understanding death as forever closing a possible future in order for one's death to be bad? This would entail that death is not at all bad for a toddler or infant (DeGrazia 1996, 237). One can have a reason to live even if one does not understand that it is a reason to live, because one does not understand what it is to live or to die. In general, it is just false that something cannot be bad for an individual unless that individual can understand why it is bad.

Cigman's view is strange in another way. She says that in order for one's death to be bad, one must believe either that death is the end or that there is an afterlife. Well, at most one of these beliefs may be true. Suppose that there is an afterlife. Now consider someone who believes there isn't one. That person does not understand death at all. He has a false belief about death. Why would having this false belief satisfy a necessary condition for death to be bad? Or consider someone who truly believes in an afterlife, but has not grasped the possibility of death being the final end. According to Cigman's view, that person's death would *not* be bad for him; but why should grasping or considering a falsehood be necessary for death to be bad? Cigman's thought might be that, whichever of these views is true, death is a "radical" transition; one need not have any particular view about what happens at death in order for one's death to be bad, but one must at least think of death as involving a radical transition. But suppose Ann falsely believes death is a radical transition into an afterlife, and Beth falsely believes it is not a radical transition at all. Why should the difference in their false beliefs about death make any difference to whether their deaths are bad for them?

I don't think premise 1 can be undermined by appeal to a sort of desire that cows do not have. Neither categorical desires, nor intrinsic desires, nor future-directed desires, nor a sophisticated understanding of the nature of death, are necessary for death to be bad.

Here is another way we might try to undermine premise 1. We might think that, although the cow's lifetime well-being level would be higher if it were to live, death is not bad for it because it is not connected in the right way to its future. In particular, we might think that cows do not have sufficient psychological connectedness over time for future goods and evils to matter to it. Future goods and evils would be, in a way, like goods and evils happening to other cows. Jeff McMahan holds a view like this, which he calls the "time-relative interest account" (TRIA) of the badness of death (2002, 105). The rough idea of McMahan's view is that when determining how bad some event is for an individual, we look not only at the goods the event deprives the victim from having, but also at how psychologically connected—via memories, desires, and the like—the person is (or would have been) between the time of the event and the time the good would have been had. If this view is correct, we might argue that death is not bad for cows because although a cow is deprived of some future goods by dying at time t, the cow's death is not bad because, when we look at how things would have gone for the cow if it hadn't died at time t, the cow at the time of the future goods would not have been psychologically connected to the cow at t.[6]

This is not *McMahan's* view about cows; he thinks death is bad for cows, just not as bad as it is for creatures that exhibit more psychological connectedness. And it seems very implausible that cows have no psychological connectedness over time. They engage in behaviors that take time to complete. Once in a while they jump over fences to escape the slaughterhouse. A creature that had no psychological connectedness over time would have a hard time doing such things. So McMahan's view is not going to help explain why death is not bad for a cow.

One might try to argue that although a cow's death is bad for it, it is not *as bad* as the death of an adult human. McMahan's view seems to have this implication, if we make some assumptions about the psychological connectedness of cows over time. We might also get this result from certain axiologies. One who likes Mill's distinction between higher and lower pleasures might claim that cows don't get any higher pleasures, so their lives are not as good as those of people, so their deaths are not so bad. One who thinks achievement or knowledge are valuable might argue that cows do not achieve or know many things, and that their lives are therefore not as good as those of people, and so their deaths are not so bad. It is a very difficult thing to compare the values of lives of members of different species. I don't know how to tell whether my dog is having a better life than mine or a worse life; I prefer my own life, but that is compatible with it being the case that his life is better *for him* than mine is *for me*. In thinking about this question, it is very difficult to be objective—almost all of us want to say that lives like ours are more valuable than the lives of very different kinds of things, at least in part because this justifies us giving preference to beings like us. Questions about interspecies comparisons of well-being are worth further exploration, but I won't explore them here.[7] Even if the death of a cow is not as bad as the death of a human, it might still be very bad for it, and we have seen no good reason to deny this so far.

3.3. LIFETIME WELL-BEING

So far there is no good reason to reject premise 1. So let us move on to premise 2. The questions here are: What features must an individual have in order to have a lifetime well-being level, and do cows have those features?[8]

In one view, to have a lifetime well-being level, one must merely have some momentary well-being levels, because an individual's lifetime well-being is just the sum of her momentary well-being levels. In this additive view, cows have lifetime well-being levels. But this view is controversial. One of the most forceful objections to the additive view is given by David Velleman.

According to Velleman, *momentary* well-being and *diachronic*, lifetime well-being are completely different things. Cows have only momentary well-being, not lifetime well-being. Lifetime well-being is not reducible to momentary well-being or even to facts about the arrangement of bits of momentary well-being (Velleman 1993). Lifetime well-being is determined (at least partly) by facts about *narrative structure*, such as whether early sacrifices pay off later in life, whether projects succeed or fail, and so on. For example, according to Velleman it is better for you to work out your marital difficulties, reconcile with your spouse, and live happily ever after, than it is to get divorced, meet someone new, and live happily ever after with that person. The reason is that you have invested a lot of effort into the previous relationship, and your life is better if those efforts pay off, even if you'd be just as well-off at every moment if you started over with someone new. Momentary well-being, on the other hand, is determined by facts about that particular moment, such as how pleased you are at that moment. Thus well-being is "radically divided" (Velleman 1993, 345). There is a component determined by momentary facts, and a component determined by narrative facts.

Velleman also holds the following thesis about intrinsic value: "unless a subject has the bare capacity, the equipment, to care about something under some conditions or other, it cannot be intrinsically good for him" (354–55). Given this thesis and our supposition about the impoverished mental capacities of cows, it follows that cows have no lifetime well-being at all. Cows cannot care about how their whole lives go; they can care at any given time only about how things go for them at that very time. So cows do not have good *lives*, only good *moments*.

These theses about well-being form the basis for Velleman's argument that death is not bad for cows, which goes as follows:

> There is no moment at which a cow can be badly off because of death, since (as Lucretius would put it) where death is, the cow is not; and if there is no moment at which a cow is harmed by death, then it cannot be harmed by death at all. A premature death does not rob the cow of the chance to accumulate more momentary well-being, since momentary well-being is not cumulable for a cow; nor can a premature death detract from the value of the cow's life as a whole, since a cow has no interest in its life as a whole, being unable to care about what sort of life it lives. Of course, a person can care about what his life story is like, and a premature death can spoil the story of his life. Hence death can harm a person but it cannot harm a cow. (1993, 357)

Velleman's argument is complicated. Let us excise the portions that concern the metaphysics of death, since these are very difficult to deal with and not necessary for Velleman's argument, as far as I can tell.[9] Here is a simplified version of his argument:

1. If death is bad for a cow, then death detracts from the lifetime well-being level of the cow. (This follows from the deprivation account of death's badness.)
2. In order for something to detract from the lifetime well-being level of a cow, it must either bring it about that the cow accumulates less momentary well-being, or it must negatively affect the cow's life story in a way the cow cares about.
3. Cows cannot accumulate well-being.
4. Cows cannot care about their life stories.

5. Death, therefore, does not detract from the lifetime well-being level of a cow.
6. Death, therefore, is not bad for a cow.

Both premises 3 and 4 are justified by the same consideration: cows have no concept of themselves existing through time; they cannot grasp their whole lives. Thus well-being does not accumulate for them over their lifetimes, and they cannot grasp their life stories. So cows do not have a lifetime well-being level at all. This claim is supported by the following principle about intrinsic value, which I'll call the "capacity to care condition" (CCC): Nothing can be intrinsically good or bad for an individual unless the individual has the capacity to care about it.

Cows can care about what happens to them at a particular moment, but since they lack the capacity to see themselves as temporally extended beings, they cannot care about extended periods of their lives, or their whole lives. Thus their whole lives lack value for them, even if moments can be good or bad for them.

The claim that extended periods of a cow's life have no intrinsic value for the cow has bizarre implications. Consider two possible futures for a cow. In one future, the cow is tortured constantly until it dies. In the other future, the cow is happy and free. Which future is better for the cow? If Velleman is right, neither future is better. The second future has better moments for the cow, but on the whole, it is no better or worse than the first future. That cannot be right.[10] Velleman claims that, given the truth of CCC, "any method of combining the values of a cow's good and bad moments will be purely arbitrary and consequently defective" (1993, 356). But could it really be arbitrary to suppose that when we combine the values of a cow's bad moments, the result is as good as, or better than, the result of combining the values of the cow's good moments? Surely not all ways of combining momentary value are equally good.

We must, then, reject premise 3; we must reject the claim that a cow's life has no intrinsic value for it, and must therefore reject CCC. But there must be some reason Velleman finds CCC to be plausible. To see what this reason might be, let us make a distinction between two ways something can be intrinsically valuable. Sometimes when something is intrinsically valuable, its value is *reducible*. It is valuable because it is made up of some parts that are intrinsically valuable. For example, the happiness of the whole class is intrinsically valuable, but its value comes from the value of the happiness of each of the individuals in the class. On the other hand, sometimes the intrinsic value of something is irreducible. Perhaps the happiness of an individual is intrinsically valuable in a way that is not reducible to the value of its parts. (Or perhaps, in the case of an extended period of happiness, its value comes from the values of its temporal parts, which themselves have irreducible intrinsic value.)[11]

Perhaps in order for something to be irreducibly intrinsically valuable for an individual, that individual must be able to care about it. Thus, in order for an individual's *whole life story* to be irreducibly good for her (such as by being a story of improvement, or success, or redemption), she must be able to contemplate her whole life. A story of redemption or overcoming obstacles does not get its value from the intrinsic values of its parts, but from *how the parts fit together* to form a whole: an early misfortune, then some efforts to overcome it that are ultimately successful. However, it does not follow that in order for something to be *reducibly* intrinsically valuable for someone, she must be able to care about it. And in fact this seems implausible. Consider the happy class: it is good for the class to be happy, even if the class does not or cannot care about its happiness, and even if each individual in the class cares (or even can care) only about his or her own well-being. Just as well-being can aggregate *across* individuals even if the aggregate of individuals cannot care about its well-being, well-being can aggregate *within* an individual even if the individual cannot care about the aggregate well-being in its life. This enables us to say that cows have lifetime well-being levels, and so a life of torture is not just as good for the cow as a happy life. Of course, there may still be one way in which cows' lives cannot be good or bad for them: their lives cannot have the kind of value that comes from narrative structure (granting Velleman's views about cow psychology). It is controversial whether narrative structure has any relevance to well-being; but even if it does, it is surely not the only component of well-being, and surely not the only component of well-being that is relevant to the badness of death.

3.4. CONCLUSION

Those who wish to argue that death is not bad for a cow have two options: (1) reject a deprivation account of death's badness in favor of some other account that entails that death is not bad for cows; or (2) claim that cows lack lifetime well-being levels, and that therefore the deprivation account cannot apply to them. Alternative accounts of death's badness are either subject to fatal counterexamples (such as "categorical desire" accounts, which entail that death is not bad for babies), or do not entail that the death of a cow is not bad for the cow (such as McMahan's "time-relative interest" account). Velleman's argument that cows do not have lifetime well-being levels is based on a principle that has very implausible implications, and that can be replaced by another very similar principle that does not have those implications. I conclude that cows can have good lives, and that death can be bad for them.

This does not show that we are morally obligated not to eat hamburgers. An important question that I have not discussed at all here is whether cows have moral *status*, or moral *rights*. It is consistent with what I have said here that although death is bad for cows, cows lack moral status, so that the badness of their deaths does not matter morally. Nevertheless, I think it cannot be argued that turning happy cows into hamburgers is not bad for the cows.

ACKNOWLEDGMENT

Versions of this paper were presented at West Virginia University in 2007 and at *Death: Its Meaning, Metaphysics, and Morality* at Newcastle University in 2011. Thanks to all those present for their helpful comments. Thanks also to an anonymous referee for helpful suggestions.

NOTES

1. For some examples of deprivation accounts, see Feldman 1992, chaps. 8–9; Broome 1999, chap. 10; Bradley 2004; and Bradley 2009, chap. 2.
2. Midgley argues that, given what we know about similarities between the biological and social characteristics of humans and other animals, "reasons must be found for *refusing* to say" that animals lack the mental states humans have (1983, 134). See DeGrazia 1996, chap. 6, for a defense of the claim that animals have desires; see Glock 2009 for a recent discussion of animal rationality.
3. Belshaw 2012, 274. For more discussion of categorical desires and alternative interpretations of the notion, see Bradley and McDaniel 2013.
4. For similar points, see Carruthers 1992, 84–85; E. Harman 2011, 730; and Norcross 2012, 469. Discussing a similar case, Belshaw says he is "not sure" that death is bad for such a person (2009, 116).
5. See DeGrazia 1996, chap. 7, for a strong case that many nonhuman animals do have a sense of time.
6. See Bradley 2009, chap. 4, for objections to the time-relative interest account. See E. Harman 2011, 733–35, for related discussion.
7. See DeGrazia 1996, chap. 8, for more detailed discussion.
8. I discuss these issues in Bradley 2009, 147–52. The discussion here is derived from the one there with some simplifications and alterations, but the main points are unchanged.
9. See Bradley 2009, chap. 3, for extensive discussion of the Lucretian argument to which Velleman refers.

10. There is an alternative line of argument that one might give. One might argue that since a cow lacks the ability to conceive of its life as a whole, cows are not temporally extended entities. Rather, at each successive moment there is a new individual, psychologically unconnected to past and future individuals but causally connected to them. There is no individual that is composed of these instantaneous cow-slices. If so, then there is no such thing as "lifetime well-being" for a cow that is distinct from momentary well-being, since no cow lives more than an instant. I take it this is not Velleman's view, so I will not evaluate it here.
11. See G. Harman 1967 and Feldman 2000 on "basic" intrinsic value.

4

The Comparative Badness for Animals of Suffering and Death

JEFF MCMAHAN

4.1. HUMANE OMNIVORISM

An increasingly common view among morally reflective people is that, whereas factory farming is objectionable because of the suffering it inflicts on animals, it is permissible to eat animals if they are reared humanely and killed with little or no pain or terror. I will refer to methods of rearing and killing animals that are designed to avoid causing significant suffering or terror as "humane rearing," and will use the label "humane omnivorism" to refer to the view that, while it is wrong to buy and eat meat produced by factory farming, it is permissible to eat meat produced through humane rearing.

Although humane rearing, when practiced scrupulously, does not cause animals to suffer, it does involve killing them quite early in their lives. Beef cattle have a natural life span of about thirty to thirty-five years, but they are normally killed at about three years of age. Pigs can live about fifteen years, but they tend to be killed at about six months, while chickens can live about eight years but are killed less than a year after birth. The reason these animals are killed when young is that it is economically wasteful to invest resources in keeping them alive after they have reached their full size.

Because humane rearing involves depriving animals of most of the life they might otherwise have, humane omnivorism seems to presuppose that depriving animals of good experiences is not morally objectionable in the way that causing them to suffer is. The assumption seems to be that although their suffering matters, their lives matter much less, and perhaps not at all.

Humane omnivorism grants that the suffering of animals matters enough to make factory farming, and supporting factory farming by eating the meat it produces, wrong. But it does not have to concede that the suffering of an animal matters as much as the equivalent suffering of a person. It might instead claim that the suffering of an animal matters less, perhaps on the ground that the suffering of beings with lower moral status matters less than the equivalent suffering of beings with higher moral status. Consider, for the sake of illustration, a version of humane omnivorism based on the artificially precise assumption that the suffering of an animal matters a tenth as much as the equivalent suffering of a person. Like all versions of humane omnivorism, this version accepts that it is permissible to kill an animal as a means of enabling people to have the greater pleasure of a certain number of meals that include meat from the animal rather than the lesser pleasure of an equivalent number of meals without meat. This is true even when, if the animal had not been killed, it could have had many more years of life without significant suffering.

Suppose that one animal would provide meat for twenty meals, and that twenty people would each get ten more units of pleasure from eating a meal with meat from the animal than they would get from otherwise similar meals without the meat, but with some substitute plant-based food of equivalent cost and nutritional benefit. We might next ask how much suffering it might be permissible to cause a *person* to experience as a means of enabling twenty other people to experience ten units of pleasure each. The answer is surely that it would be permissible to cause a person at most only a tiny amount of suffering for this reason. Given a version of humane omnivorism that accords the suffering of an animal a tenth the weight of the equivalent suffering of a person, it follows that it would be permissible to cause an animal no more than ten times this tiny amount of suffering as a means of providing the ten units of pleasure to each of ten people. Ten times a tiny amount of suffering is a small amount of suffering. So this version of humane omnivorism implies that while it would be permissible to deprive an animal of many years of life as a means of providing each of twenty people a certain amount of pleasure, it would not be permissible to cause that animal more than a small amount of suffering as a means of providing the same pleasure to the same people.

According to this version of humane omnivorism, therefore, it is worse to cause an animal to experience a small amount of suffering than it is to kill the animal, even when killing it would deprive it of many years of life without significant suffering, so that the good life it would lose would be good for it by much more than the suffering would be bad for it. Many other possible versions of humane omnivorism make similar assumptions about the relative badness of suffering and death for animals.[1] We can group these assumptions under the label "Suffering Is Worse." My aim in this article is to examine the claim that Suffering Is Worse.[2]

I will begin, in section 4.2, by considering whether there is a general moral asymmetry between suffering and happiness. In section 4.3, I sketch an account of the misfortune of death that helps to explain why the loss through death of a certain amount of good life is generally a lesser misfortune for an animal than for a person. This account, therefore, provides some support for the idea that Suffering Is Worse. Section 4.4, however, shows that the account is vulnerable to counterexamples, and section 4.5 shows that one appealing way of avoiding the counterexamples threatens to deprive the account of its distinctive virtues. In sections 4.6 through 4.9 I explain how the account's apparently implausible implications can be avoided or at least mitigated by conjoining it with either of two plausible views about the morality of causing individuals to exist. Section 4.10 then considers the implications of the two conjoined views for the morality of abortion and section 4.11 summarizes their implications for Suffering Is Worse and humane omnivorism.

4.2. POSSIBLE DEFENSES OF THE CLAIM
THAT SUFFERING IS WORSE

Many people believe that there is a general moral asymmetry between suffering and happiness for persons as well as for animals—that is, they believe that the reason not to cause or allow individuals to suffer is normally stronger than the reason to cause or allow individuals to experience a corresponding degree of happiness. Some people, for example, accept a strong form of this view, according to which pure benefits cannot on their own morally offset harms. These people think that while it can be permissible to cause a person to suffer, even without her consent, if that is necessary to prevent her from experiencing even greater suffering, it is not permissible to cause her to suffer without her consent to provide her with a benefit that would be good for her by more than the suffering would be bad (Shiffrin 1999). The claim that Suffering Is Worse in animals might simply be a corollary of this general asymmetry between causing suffering and bestowing benefits. For although we think of death as a harm, it does not involve suffering or anything intrinsically bad. It involves only the absence of pure benefits—the good experiences and activities that continued existence would have made possible.

This asymmetry is, however, difficult to reconcile with the fact that we all accept, and believe that it is rational to accept, trade-offs between suffering and happiness. People are often willing to undergo considerable suffering to achieve benefits for themselves, even when they would not suffer from the absence of those good things. If challenged, they may say that the benefits outweigh the suffering—thereby implying that the suffering and the benefits can be roughly measured on a common scale, and that the benefits would be good for them by more than the suffering would be bad for them. It therefore seems a mistake to suppose that when it is not possible to get a person's consent, it is permissible to cause him to suffer when that is necessary to prevent him from experiencing a greater harm, but not to enable him to have a greater benefit. I suspect that the explanation of our attraction to this view is that, while it is often uncontroversial that one instance of suffering is worse than another, so that one can often be confident that a harm one inflicts will be less bad than a harm one thereby prevents, there is greater diversity in what benefits people. In consequence, there is often considerable uncertainty about whether or to what extent a person would benefit from something from which some others would benefit, and thus about whether the presumed source of benefit would in fact compensate the person for the suffering that is a necessary means of bringing it about.

Another reason for skepticism about the moral asymmetry between causing suffering and providing benefits is that, although we tend to accept it in a range of cases, we reject its application to persons in exactly the kind of choice to which defenders of humane omnivorism think it applies in the case of animals—namely, choices between suffering and death. Again, if one saves a person's life, one does not prevent anything that is intrinsically bad for her; one merely enables her to have the benefits of continued existence. Yet in instances in which one can save a person's life only by causing her to suffer, we believe that it is permissible, and perhaps required, to save her, even when it is not possible to get her consent, provided that the net benefit to her in remaining alive would outweigh the suffering it would be necessary to inflict.

This challenges the view that there is a general asymmetry between causing or allowing individuals to suffer and causing or allowing them to have pure benefits. If it is permissible to cause people to suffer, even when one cannot get their consent, to enable them to continue to have benefits rather than to die, it ought to be permissible as well to cause them to suffer to enable them to have more benefits rather than continuing to live with fewer, provided the increase in benefits would outweigh the suffering.

Some defenders of humane omnivorism seem to think it unnecessary to provide arguments to show that the painless killing of an animal is justified by the benefits that people derive from eating it. They seem to assume, instead, that the painless killing of an animal that has been humanely reared requires no moral justification at all. But presumably that could be true only if the well-being or happiness of animals does not matter at all. Perhaps the best defense of that claim is the austere perfectionist view that the only pleasures that are accessible to most animals, such as eating, playing, and lying in the sun, are too "low" to matter. But while it may well be true that some higher forms of pleasure have lexical priority over certain lower ones, it is implausible to suppose that these lower pleasures have no value at all. If the lower pleasures did not matter, all that would matter in the well-being of infants and small children would be the avoidance of suffering, except insofar as their experience of the pleasures of which they are capable would be instrumental to their ability to experience higher pleasures later in life.

Someone might argue that because animals have a lower moral status, the lower pleasures do not matter in their lives, though they do matter in the lives of persons. That is, the moral status of animals is such that their lives do not matter, even though their suffering does. This, however, seems entirely ad hoc. It is hard to see what kind of rationale could be given for it. If it is compatible with their lower moral status that their suffering matters, it seems that their happiness should matter as well, at least to some extent. If, moreover, the suffering of animals mattered but their happiness did not matter in any way, there would be a moral presumption against causing or allowing any animal to exist; for even if it is not inevitable that any animal will experience suffering, it is inevitable that any animal will be at risk of suffering.

On this view, therefore, if it is morally justifiable to cause or allow an animal to exist, and thus to be exposed to the risk of suffering, that must be because of the benefits to people of its existing or the costs to people of preventing it from existing. Even more implausibly, the view also seems to imply that unless animals have a right not to be killed, which seems incompatible with a conception of their moral status that denies that their lives matter, there must be a moral presumption against allowing any animal to continue to exist. No matter how much happiness an animal's future life would contain, if it would also contain some suffering, or if the animal would be at risk of suffering, it could be permissible to allow it to continue to exist only because of the benefits to people of its existence, or because of the cost to people of killing it painlessly. But that seems false.

4.3. THE TIME-RELATIVE INTEREST ACCOUNT

The claim that Suffering Is Worse is clearly not based on the idea that there is something especially bad about the suffering of animals. It is based, rather, on the view that death is *not* especially bad for animals. Part of the explanation of this is obvious: the life that an animal loses in dying is almost always inferior in quality and quantity to the life that a person is deprived of by death. Given the orthodox, and in my view, correct view that death is bad for an individual primarily because it deprives him of further life that would have been good for him, it is natural to suppose that how great a misfortune an individual suffers in dying varies with the quality and quantity of the life he would otherwise have had. Because animals usually lose less, their deaths are usually less bad.

Many philosophers have argued that death is a misfortune for an individual in direct proportion to the amount of good life it prevents the individual from having (Feldman 1992; Bradley 2009; Broome 2013; see also Bradley, this volume). But this implies that the worst death that an individual can suffer is immediately after beginning to exist. If, for example, we begin to exist at conception, as most people seem to believe, the death of a zygote immediately after conception is the worst death anyone can suffer. Yet most people believe that if a sperm and egg were about to fuse in conception but were destroyed just prior to fusing, there would be no significant loss at all. But if that is right, it is hard to believe that if the zygote were to die immediately after the sperm and egg have fused, without experiencing even a flicker of consciousness, it would be the victim of a terrible misfortune: the worst death an individual could suffer. On any plausible view about when we begin to exist, it is intuitively implausible that death immediately after that is a greater misfortune for the individual at the time than, say, the death of a person at age twenty.

I have argued elsewhere that the extent to which death is a misfortune for an individual is a function primarily of two independent factors: (1) the amount of good life of which the individual is deprived by death and (2) the extent to which the individual at the time of death would have been psychologically connected to himself at those times in the future when the good things in his life would have occurred (McMahan 2002). On any plausible view of when we begin to exist, we come into existence either without the capacity for consciousness or in a psychologically rudimentary condition. At best we are then only very weakly psychologically related to ourselves in the future. We are unaware of having a future in prospect and thus have no future-directed desires or intentions; nor will we later have any memory of our present experience. There would be no significant difference between our dying at that point and our never having come into existence at all. But as we mature psychologically, we gradually become both more substantial as possible subjects of misfortune and more closely psychologically connected to ourselves as we will later be, if death does not intervene. Until we reach a certain level of psychological capacity, death becomes a greater misfortune as we develop, even though the amount of good life we have in prospect is steadily diminishing.

Like ourselves in the earliest moments of our lives, most animals are, throughout their lives, largely psychologically unconnected to themselves in the future. They live mostly in the present. So not only is the life they lose through death inferior in quality and quantity, but they are also only weakly related to their possible future life in the ways that matter. The magnitude of the misfortune they suffer in dying is diminished accordingly. The strength of an animal's interest in continuing to live is, one might say, discounted for psychological unconnectedness between itself at the time of death and itself at the times at which it would have had good experiences in the future. I call this the "time-relative interest account" (TRIA) of the misfortune of death. According to this account, the strength of an individual's present interest in some possible event reflects the degree to which it is rational to care for the individual's own sake now whether that event will occur. It does not necessarily reflect the way the event would affect the value of the individual's life as a whole.

The badness of causing an animal to suffer now is, by contrast, unaffected by diachronic psychological unconnectedness. Even if the suffering of an animal matters less because of its lower moral status, it may be worse for the animal to suffer now than to die now, because the extent to which its losses through death matter is steeply discounted for psychological unconnectedness. This point can be illustrated with an example:

Suffering Now. An animal has a condition that will soon kill it painlessly. One can save it, but only in a way that will cause it moderate suffering beginning shortly and continuing for a few days. It will then live for five years in a hedonically neutral state, followed by ten years of comfort, during which it will experience some of the highest forms of happiness of which it is capable.

According to the TRIA, the animal's present interest in avoiding the immediate suffering is strong, while its interest in experiencing greater happiness in the distant future is weak because it would be only very weakly psychologically connected to itself during that later time. Depending on the relative weights of the two factors (amount of happiness and degree of psychological unity), the TRIA could imply that it would be better to allow the animal to die. In this case, therefore, the TRIA supports Suffering Is Worse.

4.4. OBJECTIONS TO THE TIME-RELATIVE INTEREST ACCOUNT

The TRIA is not, however, consistent in its support for Suffering Is Worse. This is evident in a parallel example in which the temporal ordering of suffering and happiness is reversed:

> *Suffering Later.* An animal has a condition that will soon kill it painlessly. One can save it only in a way that will enable it to experience moderate happiness beginning shortly and continuing for a few days, after which it will live five years in a hedonically neutral state, followed by a few weeks of intense suffering before it finally dies. If one saves it now, there will be no opportunity to prevent it from suffering later.

If the TRIA can imply that in Suffering Now it is better to allow the animal to die, even though its future happiness would outweigh its immediate suffering, it can also imply that in Suffering Later it would be better to save the animal, even though its future suffering would outweigh its immediate happiness. The TRIA's implication in Suffering Later seems quite counterintuitive.[3] Many people may find its implication in Suffering Now implausible as well, though those committed to Suffering Is Worse should find it plausible.

Elizabeth Harman finds the TRIA's implications in both cases implausible. She argues that the TRIA is vulnerable to the following counterexample:

> Tommy is a horse with a serious illness. If the illness is not treated now and is allowed to run its course, Tommy will live an ordinary discomfort-free life for five years, but then Tommy will suffer horribly for several months and then die. If the illness is treated now, then Tommy will undergo surgery under anesthetic tomorrow. Tommy will suffer over the following two weeks, but not nearly as severely as he would five years from now. Tommy will be completely cured and will be able to live a healthy normal life for another fifteen years. (2011, 735)

Harman believes that it would be better for the horse to be treated. It is better for it to have suffering now together with greater happiness in the future (which conflicts with the TRIA's judgment in Suffering Now), rather than happiness now at the cost of greater suffering in the future (which conflicts with the TRIA's judgment in Suffering Later).

This case is not, however, a counterexample to the TRIA. Harman believes it is because she thinks that the TRIA implies that it is better not to perform the surgery. This is because it does not discount the immediate, lesser suffering that the surgery would cause but greatly discounts the greater suffering that will result if the surgery is not performed, since the horse is now only weakly psychologically connected to itself at the later time when the greater suffering would occur. But hers is a case in which the horse will exist in both possible outcomes of one's choice. Over time, its interests include a strong present interest in avoiding the immediate suffering, a weak present interest in avoiding the later suffering, and a strong future interest in avoiding the later suffering. Although the TRIA discounts interests for psychological unconnectedness, it does not claim that only present interests matter; rather, it accepts that all the interests that an individual has at all times at which it exists can be sources of reasons. Thus an agent who has no reason to care more about the horse now than in the future should choose to treat it because its *future* interest in avoiding the greater suffering will be stronger than its present interest in avoiding the immediate, lesser suffering. (There is a parallel here with a relevant difference between abortion and prenatal injury. Whereas abortion frustrates only the weak present interest of a fetus in continuing to live, prenatal injury may frustrate the strong future interests of a person.)

In Suffering Now and Suffering Later, however, one of the options is that the animal will die now. Because of this, the only interests that it has independently of the outcome of one's choice are its present interests. In Suffering Now, the horse now has a strong interest in avoiding immediate suffering but only a weaker interest (because of the weakness of the psychological connections between itself now and itself later) in experiencing greater happiness in the distant future. Its interests over time do *not* include a strong interest in experiencing greater happiness in the distant future that will exist at that later time independently of the decision about whether to save it. Similarly, in Suffering Later, it has a strong interest in having happiness now but only a weaker interest in avoiding greater suffering in the distant future. Whether it will later have a strong interest in avoiding great suffering at that future time depends on the decision about whether to save it. In short, in Harman's example, the strong interest that the horse will have in avoiding great suffering in the future is an actual, future interest that one must take into account in determining what is best for the horse. In Suffering Now, by contrast, the strong interest the animal might have in experiencing greater happiness in the future is only a *possible* interest, as is the strong interest it might have in avoiding greater suffering in the future in Suffering Later. Thus, while Harman's case is not a counter-example to the TRIA, the related case of Suffering Later is; and Suffering Now may be as well.[4]

4.5. POSSIBLE INTERESTS

Those who believe Suffering Is Worse should accept the TRIA's judgment that it is better to allow the animal to die in Suffering Now. But they and others will reject its implication that it is better to save the animal in Suffering Later, when its future life would contain significantly more suffer-ing than happiness. It seems that the relevant interests in Suffering Later include not only the animal's undiscounted present interest in experienc-ing immediate happiness and its discounted present interest in avoiding greater suffering in the future, but also the possible interest it might have later in avoiding the greater suffering when that suffering might occur. This later interest would be strong, since it would be an interest in the animal's immediate experience, and thus would not be discounted for psychological unconnectedness. If this *possible* interest can ground a reason now to pre-vent the animal from experiencing the greater suffering, that could explain and justify our belief that it would be better not to save the animal's life.

There are, however, certain problems with this. If an interest an animal would have only if we act in a certain way can give us a reason either to act or not to act in that way, then the possible interest the animal in Suffering Now might later have in experiencing happiness in the distant future could ground a strong reason now to enable it to experience that happiness. That possible interest could outweigh the animal's undiscounted present interest in avoiding the immediate suffering. But if the TRIA were to take account of possible interests in this way, it would then imply that it would be best to save the animal in Suffering Now. It would no longer support the view that Suffering Is Worse.

Many will find this implication plausible. But there is more at stake than whether Suffering Is Worse is true. As I noted earlier, the TRIA offers a plausible explanation of why death is worse for a twenty-year-old than for a fetus, even though the fetus loses more good life in dying. If, moreover, a fetus lacks the moral status that would give it a right not to be killed that is independent of the strength of its interest in continuing to live, the

TRIA's explanation of why a fetus is not greatly harmed by being killed provides the basis for an argument for the permissibility of abortion in a wide range of cases (McMahan 2002, chap. 4). Abortion can be permissible when the weak interest of the fetus in continuing to live can be outweighed by conflicting interests of the pregnant woman. But if possible interests can ground reasons now (as an animal's possible interest in not experiencing great suffering in the future seems to ground a reason not to save its life now), then a fetus's possible interests in having the good experiences of its later life can ground a reason not to kill it that is independent of any actual interest it has now. In the remainder of this paper, I will consider whether the TRIA can be supplemented in a way that enables it to address the two counterexamples (Suffering Now and Suffering Later) without forfeiting its ability to explain why a very early death is a lesser misfortune and why abortion can often be permissible.

4.6. UNCONNECTED ANIMALS AND THE ASYMMETRY

In section 4.3, I suggested that the death of an individual immediately after it begins to exist is of scarcely more significance than its not having come into existence at all would have been, given the absence of psychological connections between the individual at the time of death and itself as it would have been when the good things in its life would have occurred. I claimed this of persons, but it should be true of animals as well. Consider now an animal that lacks any degree of self-consciousness and thus, if it lives, will be psychologically unconnected to itself beyond the immediate future. Call such an animal a "psychologically unconnected animal," or "unconnected animal" for brevity. Because the relation between itself now and itself in the future is relevantly like that of an animal that has just begun to exist, it seems that it would suffer no greater misfortune in dying than an animal with an equal amount of good life in prospect that has just begun to exist. It thus seems that whether this unconnected animal continues to exist matters no more, even for its own sake, than whether an animal that would have a similar future comes into existence.[5]

The argument about abortion to which I referred depends on a similar claim. Even a fetus with the capacity for consciousness is at most only very weakly psychologically related to itself as it might later be as a person. Whether it continues to exist is therefore relevantly more like whether a person comes into existence than it is like whether an existing person, who would be strongly psychologically connected to herself in the future, continues to exist.

If the continuing to exist of an unconnected animal is relevantly like the coming into existence of a similar animal, we may be able to draw inferences about the morality of saving or killing such animals from our views about causing them to exist or preventing them from existing. And what we should believe about causing animals to exist should parallel what we should believe about causing people to exist. The view that most people seem to accept is what is sometimes called "the Asymmetry," the view that while the expectation that a person would have a life that would be bad for him grounds a moral reason not to cause or allow him to exist, the expectation that a person would have a life that would be good for him does not ground a reason to cause or allow him to exist. Suppose we understand the Asymmetry to include the view that the reason not to cause a person to exist whose life would be bad for him is as strong as the reason not to cause an existing person's life to be bad to a roughly equivalent degree. So understood, the Asymmetry can also apply to causing animals to exist.

One might wonder whether it is consistent to accept this Asymmetry but to reject, as I did earlier, a quite general asymmetry between suffering and happiness. My objection to the general asymmetry was that it seems incompatible with the permissibility of causing a person great suffering to save her life, even when it is not possible to get her consent, provided the net benefits of further life would outweigh the suffering. There would thus be an inconsistency if the Asymmetry implied that it would be impermissible to cause an unconnected animal great suffering to enable it to continue to live, when the benefits of continued life would outweigh the suffering. And I have suggested that the Asymmetry does have implications for causing or allowing unconnected animals to continue to live, on the assumption that a wholly unconnected animal's continuing to live is relevantly like a similar animal's coming into existence. But even on that assumption, the Asymmetry does not object to causing an unconnected animal to suffer to enable it to continue to live, provided the suffering would be outweighed by the net benefits of continued life. What the Asymmetry denies is that the benefits an individual would get from life provide a reason to cause that individual to exist or, in the case of an unconnected animal, a reason to cause it to continue to exist. It does not deny that these benefits can weigh against and offset the suffering the individual might also experience. One might express this by saying that the Asymmetry denies that pure benefits have "reason-giving weight," but not that they have "offsetting weight." (McMahan 2013, 5–35.)

We now have the elements of an argument that explains how an unconnected animal's possible future interests are relevant to whether it is better for it to be caused or allowed to continue to exist, or instead caused or allowed to die:

1. Whether an unconnected animal continues to exist is relevantly like whether an animal with similar prospects comes into existence.
2. The Asymmetry: if an animal's life would be intrinsically bad for it, there is a moral reason not to cause or allow it to come into existence. But if its life would be intrinsically good for it, that does not ground a reason to cause or allow it to come into existence.
3. If, therefore, an existing unconnected animal's future life would be bad for it, there is a moral reason to kill it or allow it to die. But if its future life would be good for it, that alone does not constitute a reason to cause or allow it to continue to exist.

According to this argument, there is a moral asymmetry between causing or allowing an unconnected animal to continue to exist when its future life would be bad and causing or allowing it to continue to exist when its life would be good—just as there is thought to be a parallel asymmetry in causing or allowing individuals, whether persons or animals, to exist. In both types of case, there is, at the time of choice, no individual that has a present (or future) interest in the good or bad things that might occur in what might be its future life. There are only the *possible* interests that either a possible animal or an unconnected animal might later have in avoiding suffering or experiencing happiness. (Recall that the reason why Suffering Later is an effective counterexample to the TRIA on its own is that the animal has no present or future interest in its own possible suffering in the distant future.) But assuming the Asymmetry applies in both types of case, possible interests in avoiding suffering count the way present interests in avoiding suffering count, though possible interests in experiencing happiness do not count at all—*except* in weighing against and offsetting possible interests in avoiding suffering.

That final qualification is important. We accept that the reason not to cause or allow a possible interest in avoiding suffering to exist and be frustrated can be overridden if the individual who would have that frustrated interest would be more than compensated by the existence and satisfaction of possible interests in later experiencing greater happiness. We accept, in other words, that while possible happiness that no individual has a present interest in having may provide little or no reason to cause an individual to exist or to continue to exist (that is, it may have no reason-giving weight), it can weigh against and potentially offset a comparable amount of suffering the individual might later experience (that is, it has offsetting weight).

For if this happiness had no offsetting weight, there would be a presumption against causing or allowing individuals to exist, as well as against causing or allowing unconnected animals (or unconnected fetuses) to continue to exist, even when their future lives would be, overall, worth living. I take this to be a *reductio* of the suggestion that happiness has no offsetting weight in these types of choice.

One might think that if the TRIA is conjoined with the Asymmetry, the permissibility of causing or allowing an unconnected animal to exist or to continue to live depends on the order in which the good and bad elements of its future life would occur, as it does in Suffering Now and Suffering Later. But in fact the order is irrelevant. In the absence of psychological connections between the animal now and itself in the future, the animal is equally unrelated to all parts of its future and there is no basis for discounting some relative to others.

4.7. CONNECTED ANIMALS AND WEAK ASYMMETRIES

While some animals do seem to live entirely in the present, others are psychologically connected to themselves in the future in relevant ways. We can call such animals "psychologically connected animals," or "connected animals" for brevity. Such animals are, of course, psychologically connected to themselves in the future to varying degrees. The nonhuman great apes, for example, are more closely connected than other animals that have a lower degree of self-consciousness. But even the more highly connected animals are only weakly psychologically connected to themselves in the future in comparison with the degree to which persons are connected to their future selves. Still, if the Asymmetry is correct, the comparative weakness of a connected animal's present interest in avoiding future suffering does not weaken the reason to prevent that suffering. In particular, an animal's being prevented from having a future life that would be, on balance, bad for it matters equally whether the animal is unconnected or connected—that is, whether or to what extent it now would be psychologically related to itself in the future. Indeed, preventing an animal from suffering in the future matters equally even if the animal does not at present exist and may never exist.

Matters are different, though, when we consider the prospect of future happiness. Because a connected animal would be psychologically related to itself in the future, it has a present interest in experiencing future happiness and there is thus a dimension to the badness of its loss of good life in the future that is not present in the equivalent loss that occurs when an animal fails to come into existence or when an unconnected animal dies. Its loss of good life through death thus matters more than an equivalent loss by an unconnected animal, and the stronger the psychological relations between the animal now and itself in the future would be, the more its loss matters—that is, the greater its misfortune would be in suffering that loss. Thus, in the case of a connected animal, though not in the case of an unconnected animal, the order in which the good and bad elements of its future life would occur may matter. Assuming, for example, that the strength of a connected animal's psychological connections with itself in the future would diminish with time, the loss of later good life matters less than the loss of earlier life that would be equally good.

In summary, if the Asymmetry is correct, the expectation that an animal, whether connected or unconnected, would have a life worth living provides no reason to cause it to exist. Nor does that expectation ground a reason to cause or allow an unconnected animal to *continue* to live. But the expectation that a connected animal would have a life worth living does provide some reason to cause or allow it to continue to live.

Some philosophers have, however, challenged the Asymmetry and argued instead for a *Weak Asymmetry*, according to which the expectation that an individual's life would be, on balance, intrinsically good—or good beyond some minimum level of goodness—does ground a reason to cause or allow that individual to exist, though one that is weaker than the reason not to cause or allow an individual to exist whose life would be intrinsically bad to a roughly equivalent extent (McMahan 2013; Harman 2004). Assuming that a unconnected animal's continuing to live is relevantly like the coming into existence of an animal with similar prospects, a Weak Asymmetry also implies that an unconnected animal's possible later interest in experiencing happiness provides some reason to cause or allow it to *continue* to live, though not as strong a reason as its possible later interest in avoiding a comparable amount of suffering would provide for *not* causing or allowing it to continue to live.

(One question about which I am uncertain is whether, if we accept a Weak Asymmetry, we should also accept that the reason to cause or allow a *connected* animal to continue to live is stronger than it would be if the Asymmetry were true. It may seem that the reason implied by a Weak Asymmetry to cause or allow an unconnected animal to continue to live applies as well to a connected animal, and that it combines additively with the reason to extend its life that derives from its psychological connectedness to its future selves. But, if this is true, it seems that the reason to extend the life of a *person* should also be stronger if a Weak Asymmetry is true than it would be if the Asymmetry were true. But it does not seem that the strength of the reason not to kill people, or to save them, depends upon whether there is a moral reason to cause people to exist if their lives would be worth living. I leave this issue open here.)

4.8. IMPLICATIONS OF THE TRIA COMBINED WITH THE ASYMMETRY

Suppose we combine either the Asymmetry or a Weak Asymmetry with the TRIA. The resulting views have mostly plausible implications for the cases that challenged the TRIA on its own—namely, Suffering Now and Suffering Later. Let us consider instances of Suffering Now and Suffering Later involving different kinds of animal and apply the different principles to them. Consider first the conjunction of the TRIA and the Asymmetry.

1. *Suffering Now with an unconnected animal.* The animal has little present interest in avoiding the imminent suffering, and no present interest in experiencing greater happiness in the distant future. According to the Asymmetry, the suffering it would experience grounds a reason to prevent if from continuing to live, while the greater happiness it might experience provides no reason to enable it to continue to live. It may therefore seem that it would be better not to save it. While this is consistent with the view that Suffering Is Worse, it is intuitively implausible. But while the happiness the animal might experience has no reason-giving weight, it nevertheless has offsetting weight vis-à-vis the suffering that would occur earlier. Because there are no psychological relations between the animal now and itself at any time beyond the immediate future, the offsetting weight of the possible future happiness cannot be discounted for psychological unconnectedness. The temporal order of the suffering and happiness does not affect their respective weights. Hence the prospect of the greater happiness offsets the lesser suffering, making it permissible to save the animal. But because the happiness has no reason-giving weight, it is also permissible to allow the animal to die, just as it would be permissible not to cause a similar animal to exist. This seems a plausible result.

2. *Suffering Later with an unconnected animal.* The animal has little present interest in experiencing the immediate happiness, and no present interest in avoiding the greater suffering later. But according to the Asymmetry, there is a reason to prevent an individual from suffering even if it has no present interest in avoiding it. This reason is as strong as the reason to prevent an existing individual of the same sort from experiencing equivalent suffering now. It is therefore better, according to this combination of views, not to save the animal, which is intuitively the correct result.

3. *Suffering Now with a connected animal.* This seems to me the most problematic case. In cases involving causing individuals to exist, or causing unconnected

animals to continue to live, there can be no discounting of the offsetting weight of future happiness for psychological unconnectedness. But a connected animal would be more closely psychologically connected to itself in the immediate future than in the distant future. It therefore has at least a moderately strong present interest in avoiding the immediate suffering but only a very weak present interest in having the greater happiness in the distant future.

According to the Asymmetry, there is no reason independent of the animal's present interests to enable it to have the greater happiness for its own sake. The question is whether the greater happiness in the distant future can offset the lesser suffering in the immediate future. It seems, however, that the offsetting weight that the animal's possible later happiness has vis-à-vis its possible immediate suffering ought to be discounted for the weakness of the psychological connections between itself now and itself later. Even though the happiness it might experience later is substantially greater than the suffering it might endure now, it may be that the animal cannot be compensated for lesser suffering in the immediate future by the greater happiness of a later individual that would be so weakly psychologically related to itself now, even though that individual would be itself. It may be, in other words, that the animal's steeply discounted interest in experiencing the greater happiness cannot outweigh its only slightly discounted interest in avoiding the lesser suffering. If so, it may be better to allow the animal to die, even though its future life would be, on the whole, worth living. This, I concede, seems implausible, but it is not highly implausible, and it does provide limited support for Suffering Is Worse, which many people accept.

4. *Suffering Later with a connected animal.* The animal has a moderately strong present interest in experiencing the immediate happiness, but only a very weak present interest in avoiding the distant greater suffering. But according to the Asymmetry, there is a reason to prevent the suffering that is independent of any present interest in its avoidance. Because the suffering outweighs the happiness, it is better to allow the animal to die.

4.9. IMPLICATIONS OF THE TRIA COMBINED WITH A WEAK ASYMMETRY

Next consider the implications for the same four cases of a combination of the TRIA and a Weak Asymmetry.

5. *Suffering Now with an unconnected animal.* In the case of an unconnected animal, the order in which good and bad experiences occur makes no difference, because there are no psychological relations that could be either stronger or

weaker over time. According to any Weak Asymmetry as understood here, the suffering in the immediate future counts fully, independently of any actual interest in its avoidance, while the prospect of greater happiness in the distant future provides only a weaker reason to enable the animal to continue to live. But even if a Weak Asymmetry gives far greater weight to the avoidance of suffering than to the experience of happiness, these are reason-giving weights, so that all the Weak Asymmetry implies is that there may be no duty, all things considered, to save the animal *for its own sake now*. It allows that the greater happiness has sufficient offsetting weight to make it permissible to save the animal.

6. *Suffering Later with an unconnected animal*. A Weak Asymmetry implies that the happiness in the immediate future provides some reason to enable the animal to continue to live. But because the greater suffering is not discounted for psychological unconnectedness over time, and because that suffering has full offsetting weight vis-à-vis the happiness, it is better to allow the animal to die.

7. *Suffering Now with a connected animal*. The remarks under number 3 above apply here as well. Whether it would make a difference if a Weak Asymmetry were true depends on the question raised parenthetically at the end of section 4.7—namely, whether the reason implied by a Weak Asymmetry to cause an individual to exist if its life would be worth living combines additively with a connected animal's psychological connectedness to itself in the future to strengthen the overall reason to cause or allow it to continue to live. If it does, it is more likely that the combined TRIA and Weak Asymmetry will imply that it is better to cause the animal to survive by treating it. And that would be a reason to accept a Weak Asymmetry rather than the Asymmetry. But, as I noted, it is uncertain whether a reason to cause individuals to exist would then be an independent reason to keep them in existence in addition to the familiar reasons concerned with interests and rights to cause or allow connected animals and persons to continue to live.

8. *Suffering Later with a connected animal*. Assuming a Weak Asymmetry, the animal's present interest in experiencing the immediate happiness provides a reason to enable it to continue to live that is stronger than the reason to enable an unconnected animal to continue to live. But we are assuming that a Weak Asymmetry also implies that the reason to prevent the possible later suffering is as strong as the reason to prevent the animal from experiencing equal suffering now. So it is better to allow the animal to die.

These implications of the two views for the four possible cases are mainly plausible, with 3 and 7 as exceptions—though those who find Suffering Is Worse plausible may welcome the implications described in 3 and 7. What seemed to be implausible implications of the TRIA for unconnected animals in Suffering Now are blocked by appeal to the distinction between reason-giving weight and offsetting weight, while the TRIA's implausible implications for Suffering Later are blocked by combining it with either the Asymmetry or a Weak Asymmetry. There are, of course, objections to these latter views, but abandoning them in favor of full Symmetry between causing miserable individuals to exist and causing well-off individuals to exist would have highly counterintuitive implications (McMahan 2013).

4.10. ABORTION

In this short section I will consider the implications of the two views (the TRIA combined with the Asymmetry and the TRIA combined with a Weak Asymmetry) for the morality of abortion. One could plausibly argue that a human fetus is relevantly like an unconnected animal, in that it is wholly psychologically unrelated to itself in the future. But I will assume that a fetus is instead relevantly like a connected animal, and therefore weakly psychologically related to itself as a person. This assumption is less favorable to the view that abortion is often permissible. If a fetus is relevantly like a connected animal, then both views imply that if a fetus faces the prospect described in Suffering Now, it could be better to allow it to die. That, I have conceded, is implausible, though perhaps not highly implausible.

It is, however, improbable that a fetus could face a prospect such as that in Suffering Now. The typical expectation is that a fetus's future life would be overall worth living, with good and bad elements more or less evenly distributed throughout. That is, good and bad experiences normally alternate, with more good experiences than bad. Suppose we accept the TRIA and the Asymmetry. Even though the suffering the fetus would experience at various times in its life counts against allowing it to continue to live, the greater happiness it would experience *around the same later times* outweighs that suffering and thus offsets the reason to prevent it from continuing to live. This is so despite the fact that the fetus now would be only weakly related to itself during the periods of happiness. Suppose that a fetus would later, as a person, experience suffering at a certain time, but would also experience happiness shortly before and shortly after the suffering. Given that the fetus now would be psychologically related to itself to much the same degree at all three times, both the happiness and the suffering should be discounted to the same degree in the determination of their offsetting weights. Thus, if the experiences of happiness would together be greater in amount than the suffering, they would offset it. If, therefore, the fetus's life would be worth living overall, it would be permissible to allow it to continue to live. But given the Asymmetry and the fact that the fetus would be only very weakly psychologically related to itself as a person, the happiness it might later experience has very little reason-giving weight—that is, it provides only a very weak reason to cause or allow the fetus to continue to live. The fetus's weak interest in continuing to live could therefore be outweighed by the interests of the pregnant woman.

It is, of course, insufficient to deal with a counterexample to show that the case is unlikely to arise in practice. But there is more that can be said. The TRIA asserts that the misfortune of death is a function of two factors: the value of the life lost and the degree to which the victim would have been psychologically connected to that life. It does not assert what the relative weights of the two factors are. It is therefore possible that, even though a fetus would, in a case like Suffering Now, be much more closely psychologically connected to itself in the near future than in the distant future, the sheer magnitude of the value of the later life it would lose in dying could outweigh the immediate suffering. This is far more likely to be true in the case of a fetus than in the case of a connected animal because persons generally have lives of much greater quality and duration than those of animals. It is also possible that, even though the fetus's present interest in having greater happiness later outweighs its interest in avoiding suffering now, its overall interest in continuing to live is still sufficiently weak to be outweighed by the interest of the pregnant woman in having an abortion.

4.11. CONCLUSION

In the last few sections I have sought to defend the TRIA against objections by combining it with either the Asymmetry or a Weak Asymmetry. If it can be defended in this way, it may provide some support for humane omnivorism, as it implies that death is a lesser misfortune for animals than for persons, not only because their future lives would be less good, but also because of their lesser psychological connectedness to themselves in the future. In particular, if the TRIA is correct, the painless killing of wholly unconnected animals is relevantly like preventing animals with comparable prospects from coming into existence, which few believe would be wrong.

The reason not to kill connected animals is different and stronger, but less strong than the reason not to kill persons, at least in part because their loss of a less valuable future to which they would be less closely connected is a lesser misfortune. Still, even if the TRIA is true, whether humane omnivorism is permissible depends on several considerations. One is whether animals used for food are unconnected or connected. This is, of course, an empirical matter, but I suspect that only the really lower forms of animal are wholly unconnected. Most of the animals that could be humanely reared for human consumption are connected to varying degrees: pigs more than cows and cows more than chickens. And the relevant connections are stronger between the animal now and itself in the near future than between itself now and itself in the further future. This makes it possible that the satisfaction that a connected animal would get just from eating over, say, the next month could outweigh the difference in pleasure that people would derive from eating the meat of this animal rather than eating vegetarian meals. This too is an empirical question but again it seems doubtful that in most cases the difference in human pleasure would outweigh the loss of animal pleasure. If, moreover, we reject the TRIA, so that none of the happiness that an animal might experience over many further years of life would be discounted, the great majority of killings required by humane omnivorism would inflict on the animal victim a loss that would be vastly greater than the benefits it would provide to "humane" omnivores.

ACKNOWLEDGMENT

For written comments on an earlier draft, I am deeply grateful to Roger Crisp, Maxime Lupoutre, Theron Pummer, Tatjana Višak, and especially Derek Parfit.

NOTES

1. An alternative basis for humane omnivorism is an argument that Peter Singer calls the "replaceability argument." See Singer 2011, 104–119. I will not discuss that argument here, though I have done so in McMahan 2008, 66–76. See also Višak and Kagan, this volume.
2. I have written material relevant to this issue before, first in McMahan 2002, 182, 195–98, 199–203, 229–30, 474–75, 487–93.
3. I discussed this problem in McMahan 2002, chap. 5, sec. 2.4, but without achieving any resolution.
4. Harman might argue that, although her example is presented as a choice between doing and not doing the surgery, so that the horse's later interest in avoiding suffering is a *future* interest relative to that choice, there is nothing that excludes killing the horse as an option. If we explicitly include that among the options, the horse's later interest becomes a *possible* interest and her example then has the right form for a counterexample to the TRIA.
5. This claim is similar to that which is the basis for Peter Singer's suggestion that non-self-conscious animals may be "replaceable." See Višak and Kagan, this volume.

5

Animal Interests

STEVEN LUPER

Most people have inconsistent views about the nature of animals and their interests. Many animals fare well or ill, and are benefitted or harmed, because of what happens to them. Or so most of us think, and strong empirical evidence backs us up. Most of us also think that we are not animals. (We will acknowledge that human animals exist. How could we not? They are everywhere we go! One of them has eyes scanning this sentence. Still, they are one thing, and we are another; perhaps we are inside of these animals, and perhaps we overlap with them—only a trained metaphysician can detect where the one creature leaves off and the other begins.) However, one or the other view must be false, for, as I will show later, if we are not animals, no animal has interests. So if we insist that we are not animals, we must argue as follows:

1. We are not animals.
2. If we are not animals, then no animals have interests.
3. No animals, therefore, have interests.

Call this the zombie argument, since its conclusion suggests that animals are rather like zombies (philosophical zombies, not the kind that emerge at night to eat brains). It is a strong claim, for if animals lack interests, it does not matter how we treat them, except insofar as what we do to them bears on *us* (and perhaps on certain aesthetic values).

Despite the zombie argument, we need not give up the view that animals have interests. However, if we don't give it up, we will have to turn the zombie argument on its head—and accept the consequences. We will have to argue as follows:

1. Some animals have interests.
2. If we are not animals, then no animals have interests.
3. We are, therefore, animals.

In this paper I will attempt to do two things: first, I want to support the claim that is taken for granted by both of these arguments: either no animals have interests or we are animals. Second, I hope to reassure those who think that animals have interests which we must respect. The zombie argument is not as powerful as it may appear to be. The inverted argument is at least as strong.

I will begin with a rough account of interests and the features that an individual (any sort of individual) must have in order to have interests. This will help me show that various sorts of organisms, including typical human animals, have interests. Then I will sketch some accounts of what you and I are—what sorts of creatures we are—and how we are related to animals. These accounts, which are widely accepted today, say that all of us overlap with animals, but that we ourselves are not animals. This sketch sets the stage for an explanation of why animals lack interests if we are not animals.

5.1. INTERESTS

Not everything has interests. In a sense, lubrication is good for ships, and water is good for plants. Oil makes a ship operate smoothly, and water helps a plant grow; the oil and the water make the ship and the plant good exemplars of their kind. However, battleships and begonias lack interests. Nothing boosts or lowers their well-being, and it is impossible for anything to benefit or harm them. The concept of well-being just does not apply to them.

By contrast, you and I and other creatures have interests that are easily affected by the world around us. Our interests may be analyzed in terms of our welfare or well-being, which, in turn, may be analyzed in terms of those things that are intrinsically good for us, such as pleasure, and those things that are intrinsically bad for us, such as pain. My own welfare level during some period of time is determined by my portion, during that time, of those things that are intrinsically good for me together with my portion, during that time, of those things that are intrinsically bad for me. The more goods I accrue during that time, the higher my overall welfare level.

Likewise, the more evils I incur, the lower my welfare level. My *lifetime* welfare level is determined by assigning a value to the intrinsic goods I gain over the course of my life, which (assuming the goods are at least roughly commensurable) can be represented as a positive number, then assigning a value to the intrinsic evils I accrue, which can be represented as a negative number, and summing the two. Taking the notion of my lifetime welfare level for granted, we can say that an event or state of affairs is in (or against) my interests, or overall good (or bad) for me, just if, and to the extent that, it makes my lifetime welfare level higher (or lower) than it would otherwise have been. The same, mutatis mutandis, goes for all creatures. Call this view "comparativism." (Perhaps I should add that it is not always prudent for me to bring about an event that will benefit me, since there might well be an alternative that would benefit me more. Prudentially speaking, I would choose the alternative that boosts my lifetime welfare level as much as possible.)

The comparativist account of interests does not commit us to any particular view concerning which things are intrinsically good and which are intrinsically bad (nor to the assumption that all goods are cumulative; for discussion, see Velleman 1991 and Luper 2012). It seems plausible to say, as hedonists might, that pleasure is intrinsically good, while pain (not mere nociception), especially suffering, is intrinsically bad, for any creature that accrues it. (On the distinction between pain, suffering, and nociception, see, e.g., DeGrazia 1991.) It also seems plausible to say, as preferentialists might, that the fulfillment of certain desires is intrinsically good for any creature that has them, while the thwarting of those desires is intrinsically bad for that creature, even though this cannot be true of all desires. Both claims are controversial, of course; hedonists say that pleasure is the *only* good, while preferentialists restrict the good to desire fulfillment. Pluralists are likely to say that both are good, and they might also say that there are intrinsic goods beyond pleasure and desire fulfillment. In what follows I take no stand on which of the major accounts of the good is correct. Instead, I will assume something quite modest—something consistent with hedonism, preferentialism, and pluralism—namely, that pleasure *or* the fulfillment of certain desires is good for any creature.

According to comparativism, an event can be in our interests even though it has no impact on our level of welfare at the time it occurs. In order to illustrate the point, let me introduce a bit of terminology. Let us say that an individual is *responsive* at a time just if that individual has the features required for faring well or ill. For example, suppose, that pleasure (or pain) is one of the goods (or evils) that make for welfare. Then anhedonic people, who are incapable of pleasure or pain, are at least partially unresponsive.

If desire fulfillment also makes for faring well, then anhedonic people who lack the capacity to desire would be even less responsive. Quite possibly, some people who are responsive at some times are wholly unresponsive at others. If, during some time period, they are unable to accrue any of the goods and evils that make for welfare, they are wholly unresponsive during that time. However, events that occur while such individuals are unresponsive could still benefit or harm those individuals. They might accrue the harm or benefit during some other time, when they are responsive. For example, if I come to be unresponsive due to some injury, and while unresponsive I am given a treatment that later restores my responsiveness, after which I enjoy many good years of life, the treatment would have benefitted me even though it took place while I was unresponsive. It is also possible for events that occur while I am unresponsive to benefit me even though there is *no* time at which I accrue the benefit. Suppose I become unresponsive one week and die the next, but had I not died I would have been cured, only to live on in unmitigated misery for many years. In that case my death benefits me; because of it, I do not endure many years of relentless suffering; nevertheless, there is no (period of) time at which, because of it I am better off. (For more about welfare, see Feldman 1991; Luper 2009a, 2014a.)

5.2. ANIMAL INTERESTS

Using the comparativist account, we can reach some defensible conclusions about the interests of animals. Recall that, according to comparativism, what boosts (or lowers) an individual's overall welfare level is good (or bad) for her, and her welfare level is determined by the intrinsic goods and evils she accrues. It follows that an individual has interests if she can accrue intrinsic goods or evils. I mentioned some plausible candidates for things that are intrinsically good, such as pleasure and desire fulfillment, and intrinsically evil, such as pain and thwarted desires. Since the matter is controversial, I have assumed only that an individual's pleasure or desire fulfillment is intrinsically good for that individual. Given that, I can establish that some animals have interests by showing that some are capable of pleasure (or pain) and some may fulfill certain desires (or have these thwarted). Do any animals qualify?

We can be confident that some animals do not. The clearest examples would be animals such as sponges that have no nervous system whatever. But animals with really primitive nervous systems probably do not qualify either. Take *C. elegans*, a tiny roundworm, just visible to the naked eye, which is among the simplest of animals. It has been studied extensively (its genome has been sequenced, the development of each of its cells has been documented, and its nervous system, composed of 302 neurons with 7,000 synapses, has been completely mapped out). It has proven to be useful in the study of simple forms of learning (Lau et al. 2013), but to my knowledge no scientist has located any features of *C. elegans* that would indicate the ability to experience pleasure or pain.

Might the line be drawn at the vertebrates? Perhaps (see Smith and Boyd 1991), but there is evidence that some invertebrates, too, can feel pleasure or pain. Animals with surprisingly small brains may well prove to be sentient. As Lars Chittka and Larry Niven (2009) argue, smaller brains need not be less cognitively sophisticated than larger brains. (An animal's brain size tends to be associated with its body mass; thus, for example, sperm whales and elephants have much larger brains than human animals). The brains of insects such as honey bees may be far more sophisticated than might be expected from their size. There is also evidence that some insects have reward centers in their brains. According to Andrew Barron and colleagues (2009), honey bees meet this description. This claim is based, in part, on the observation that cocaine influences the dance behavior of bees via its effect on reward centers in their brains (compare Barron, Søvik, and Cornish 2010 and Perry and Barron 2013).

In any case, there are very good grounds for the claim that many animals have interests. (For extended discussions of animal sentience, see Varner 2012 and Linzey 2009.) It is obvious to anyone who has read about the self-stimulation experiments that were done in the 1950s (e.g. Olds 1954, 1955, 1958) that rats can feel pleasure (they will continue pushing a lever wired to the pleasure center in their brains until they collapse with exhaustion). And it seems obvious that other mammals, such as dogs and cats, desire things and know pleasure and pain. When injured, as she once was during an encounter with a porcupine, my dog whines and yelps; it is hard not to believe that she is evincing pain. After she paws at the back door she usually goes outside; it is hard not to believe that, by scratching at the door, she is expressing (roughly) the desire to go outside. This sort of evidence is available to anyone who cares to notice it. And few doubt that more sophisticated animals such as dolphins and chimpanzees form and fulfill desires.

Now consider human beings, animals of the species *Homo sapiens*: don't they have the apparatus that is necessary to feel pleasure and pain and to form desires? (Isn't this true of most of them most of the time, anyway?) Isn't it obvious that they have interests? All of us seem especially well positioned to answer these questions, since we seem capable of judging firsthand, so to speak. I can personally attest to feeling pain when I stub

a toe, to feeling pleasure when I drink coffee in the morning, and to forming and fulfilling the desire to write this paper. What is more, when I fulfill
desires or feel pain, there certainly seems to be an animal close at hand
incurring the same pain and fulfilling the same desires. Unlike a sponge,
this animal has a brain that enables it to enjoy pleasure and to form desires.

In the previous paragraph, I came perilously close to saying that I *am* an
animal. Let me walk that back. I will discuss some views about what I am
(what we are) in the next section. My conclusion so far is simply that some
animals, among them human animals, have interests.

5.3. PERSIMALS

In order to facilitate the discussion of what kind of thing you and I are,
let me introduce some terminology. Let the term "persimal" refer to the
kind of thing that, fundamentally, we are, *whatever* that turns out to be.
Thus, by stipulation, we are persimals, and the question *What are we?* can
be restated: What are persimals?

One possible answer is rather simple: persimals are animals—each of
us is identical to a human animal. This account of persimalhood is called
"animalism" (Snowdon 1990; Olson 1997, 2007). There are many alternatives to animalism; I will not mention all of them. I will instead focus on a
group of accounts that might be called "mentalist." A mentalist account is
one that says that only some mental (or psychological) attribute or attributes determines what persimalhood consists in. Consider a couple of
examples. One I will call "personism"; this is the position (taken by John
Locke [1975, book 2, chap. 27] and Derek Parfit [1984, 2012], among others) that persimals are *persons*, where personhood consists in having the
capacity for self-awareness. The second I will call "mindism"; this is the
view (defended by Jeff McMahan [2002]) that persimals are *minds*, where
being a mind consists in having the capacity for consciousness. Assuming
that persimals are substances (as opposed to attributes, say), then mentalism suggests that we are mental substances: we are substances of a
kind that is determined by some mental attribute, such as consciousness
or self-awareness. This is not to imply that persimals are wholly mental entities—that they have no physical features, as Descartes thought.
Perhaps it is physically impossible for the constitutive mental feature of
persimals, whether it is the capacity for consciousness or self-awareness
or something else, to exist in separation from certain physical properties. In that case, while we can *imagine* purely mental persimals, there will
never actually be any.

Given any version of mentalism, we are not animals. Yet each of us is related to a human animal in an intimate way. How is a matter of controversy. One view is that each of us is a part of an animal, a part that is itself a mental substance. Perhaps, for example, persimals are brains or specific parts of brains. According to McMahan (2002) we are the part of the brain that supplies the capacity for consciousness. In this approach, a human animal lives a while, and then develops a part which is itself a persimal. The animal and its persimal part remain in existence for a while, and then they die simultaneously or in succession.

Another view is that each of us is realized in an animal, much in the way that a statue is realized in a bit of bronze yet not identical to that bronze. Maybe we are realized in an entire animal, so that, while we exist, its boundaries and ours coincide. Or perhaps we are realized in only part of that animal—maybe in some part of its brain.

If the mentalist account of persimals and their relationship to human animals is true, then something similar must be true of many other kinds of animals. If persimals are those mental substances whose essential feature is the capacity for self-awareness, as personism says, then every self-aware animal, including chimpanzees and dolphins, hosts a persimal. If persimals are those mental substances whose essential feature is the capacity for consciousness, as mindism says, then every sentient animal hosts a persimal. A dog is an animal, but the mind within it is a persimal: *it* is no more an animal than you or I. There are far more creatures in the world than biologists have noticed.

5.4. PERSIMAL INTERESTS

At this point it should be possible to shed light on how the interests of us persimals are related to those of animals. If I *am* my animal, my interests are identical to his. The situation is less straightforward if animalism is false. If we are mental substances of some sort, do we have the apparatus that we need to form and fulfill desires, and to accrue pleasure and pain? Are we responsive? Does what happens to us matter to us? Presumably, the answer to these questions is "yes." In that case, it is possible to show that *our* interests may differ from and even clash with the interests of our animal hosts. I will do that later. However, it is not clear that the answer to these questions really *is* "yes." The problem is that we might turn out to be mental substances that are unresponsive. We might have no interests.

Consider a hypothetical creature that I will call the *schuman* (classification: *Schomo schapiens*). Schumans are animals whose brains are much like human brains, but in schuman brains there is a clear differentiation among parts that supply certain key capacities: consciousness and awareness, desiring, enjoyment and suffering, recall, and belief and reason. Schumans are able to form and retain desires using one module of their brain, and they are aware of these desires through the operation of another module, the module that supplies consciousness, when the latter gains access to the contents of the former. Schumans remember things using the recall module and become aware of their memories when their consciousness module gains access to the contents of their recall module. A similar relationship exists between their consciousness and the modules where their beliefs are formed and where their reasoning occurs. Certain states associated with pleasure and pain occur in the module concerned with feeling; pleasure and pain are experienced when consciousness gains access to these states. The contents of the modules are, of course, distinct from, and not essential to, the modules themselves; for example, the recall module is distinct from the particular memories it retains. If one module of the schuman brain fails, the others may well continue to function. For example, the loss of the recall module would leave a schuman that is still conscious and that still desires and feels but that is utterly unable to remember anything.

We can use schumans to illustrate how persimals could lack interests if mindism or personism is true. First, however, I will need to add an assumption about the boundaries of the minds or persons that are hosted within animals: let us stipulate that minds include all of and only the part of the animal that supplies the capacity for consciousness, while persons include all of and only the part that supplies the capacity for self-awareness. Based on these assumptions, the persimals in schumans lack the apparatus of responsiveness. That apparatus is mostly located outside of the persimals themselves—elsewhere within the schuman brain. The persimals are, or are wholly within, the module that supplies consciousness and awareness. Given the layout of the schuman brain, their persimals only become aware of desires by interacting with the module that forms desires; they could not know desire at all if detached from that module. The same would be true if they were detached from the module involved in the formation of pleasure and pain. It is plausible to say that *schumans* have interests, for they are not only capable of consciousness and self-awareness, they are responsive, too. *They* have all of the relevant apparatus. But their persimals have no interests. Schumans also have what it takes to be considered moral agents: the relevant desires and beliefs. Their persimals do not.

Although schumans are merely hypothetical creatures, the point just made about them may apply to human animals as well. Unlike the schuman brain, the human brain is not so neatly differentiated, but it may still turn out that the regions of the human brain that supply the capacities involved in responsiveness are located largely outside of the region that supplies the capacity for consciousness and self-awareness. In that case, what was said about schumans may apply to human beings as well: their persimals lack interests. If we persimals lack interests altogether, it goes without saying that the interests of our animal hosts cannot clash with ours.

Perhaps, however, we persimals are mental substances whose boundaries are not as narrow as I have portrayed them, in which case we may well be responsive. Even if we are minds or persons, and the capacities for consciousness and self-awareness are supplied entirely by certain parts of the brain, it is possible to deny that our boundaries coincide with these areas. Mentalists might instead stipulate that persimals include any and all components of animals that are suitably related to the parts that provide consciousness or self-awareness. Perhaps persimals include all parts of the brain involved in responsiveness. Alternatively, mentalists might deny that persimals are identical with any parts of animals. They might instead say that persimals are *realized in* animals, whether in whole or in part. They might then stipulate that if the capacities involved in responsiveness are realized in an animal, then these are part of the persimal that is realized in that animal. I will not attempt to fill in the details of these strategies for ensuring that persimals themselves have interests. Instead, for the sake of argument, I will assume that one or the other strategy can be made to succeed.

5.5. THE CLASH OF INTERESTS

Suppose that we combine this assumption—persimals have interests—with a conclusion we reached earlier; namely, that human animals have interests. If both of these claims are true, then we must accept something that is, I take it, absurd: that an animal and its persimal can have distinct interests that can clash.

The argument is quite straightforward. If an animal and its persimal have interests, presumably the very same mechanisms make this possible for each of them. I, who (according to mentalism) am a persimal, hereby stipulate that my name, Steven Luper, refers to me, the persimal, and not to my animal host. As for the latter, call him Ishmael. Presumably, the apparatus that enables one of us to be responsive enables the other to be responsive. Both of us are responsive in virtue of the fact that we accrue pleasure or pain, or fulfilled or thwarted desires, or any other things there might be that are intrinsically good or bad for us. Unless something quite unlikely occurs, we will accrue the very same goods and evils, and Ishmael's lifetime welfare level will be exactly the same as mine (I consider some reservations about this view in Luper 2014b). Nevertheless, it is easy to imagine circumstances in which that would not occur. Consequently, it is easy to imagine things that would be good for me but very bad for him.

Let us say that Luper grows tired of his unremarkable looks, and considers having the latest in cosmetic surgery: a brain transplant. In this procedure, the brain is transplanted into an attractive new body whose own brain has been removed. In my case, let's make it the body of a handsome movie star (a former navy seal) whose own brain has died. Suppose that, for one reason or other, the procedure will be fatal to Ishmael, but if all goes well, Luper, the mental substance, the persimal, will sport an attractive new body. After all, he is part of or is realized in Ishmael's brain, which survives the procedure after being moved to the new body. By contrast, Ishmael will be dead. Clearly the surgery will be in Luper's interests, assuming that, as one of the beautiful people, he would fare better over his remaining life than he would otherwise have done, but it is equally clear that the procedure is very bad for Ishmael, assuming that he would have lived relatively happily over his remaining life had the procedure not been performed.

Perhaps you are tempted to say that Ishmael's interests must just be Luper's, because Luper always does the business of having interests for both of us. (For the cognoscenti: compare McMahan's [2002] and Parfit's [2012] claim, in response to Olson's "too many thinkers" problem, that persimals do the thinking for their animals.) Let me help you resist temptation. If Ishmael's interests were whatever Luper's are, then it would be in Ishmael's interests for the transplant to take place, since this is what is best for Luper. But in all likelihood the transplant is in Luper's interests because of goods that Luper will accrue after he and Ishmael part ways. Since Ishmael will be dead then, he cannot accrue these. And not only is it in Luper's interests on the day of the operation to accrue these goods, it will still be in his interests to accrue them long *afterward*. Alas, these acquisitions will come too late to benefit poor Ishmael. (Picture Luper with his arm outstretched, holding Ishmael's skull aloft, muttering "he hath borne me on his back a thousand times. . . .") So his interests and Luper's cannot possibly be identical. (Let me anticipate a particularly esoteric rebuttal that will occur to those who accept Parfit's claim that "what matters" can be separated from identity. Parfit bases this claim, his detachment thesis, on an assumption about a case in which I undergo "division," much like an amoeba undergoing reproduction, giving rise to two individuals, Lefty and Righty. He assumes that the good lives which Lefty and Righty attain would be as good for me, or very nearly so, as my normal survival would be.

Appealing to the detachment thesis, one might argue that in the transplantation case, the good life Luper accrues post-op benefits Ishmael and Luper alike: even though Ishmael does not survive, he has everything that matters in survival. However, Parfit's thesis is false [Luper 2009a, 167]. As comparativism implies, an individual's interests are mine only if I am that individual, and the accrual of a good is in my interests only if *I* accrue it or am benefitted indirectly, as when my dog's pleasure brings me happiness. Since I do not survive Parfit-style division I cannot accrue Lefty's or Righty's goods, so at best these accruals benefit me only indirectly, and may well not benefit me at all. Hence division is not nearly as good as ordinary survival.)

You might instead be tempted to say that the possibility of clashes between Luper's and Ishmael's interests is real but unimportant, since one of the two parties is a mere animal. Who, after all, cares about animals? However, anyone who is inclined to wave away the problem on these grounds has overlooked something: if persimals and human animals do indeed have their own interests, then these interests are equally significant. Recall that under normal circumstances a human animal's lifetime welfare level will exactly match that of its persimal. So, as a rule, killing human animals harms them just as badly as killing *persimals* harms *them*. Consider, too, that persimals like me generally become moral agents, and have a certain sort of significance as a result, in part by grasping certain moral principles, adopting certain preferences, and recognizing the force of certain duties. But so do human animals like Ishmael. Hence, as far as I can tell, his interests are on a par with my own, and since any injury that harms him harms me just as much, it is twice as bad as it might seem! (Still inclined to bite the bullet? Then riddle me this: if Luper goes ahead with the cosmetic transplant, can he be charged with murder? Might we instead—or in addition—say that Ishmael has opted to kill himself? Suppose the example is modified as follows: Luper is considering the procedure on the grounds that Ishmael has a slow-growing but terminal form of cancer, so that if Ishmael lives on, he will live well for five more years, then quickly die. The procedure is still bad for Ishmael and good for Luper, but can Luper now justify the procedure on the grounds that it is necessary to save his life? How selfish can you get? Do these questions really have answers?)

To be sure, I have illustrated the possible clash between the interests of persimals and their animal hosts using a thought experiment involving a medical procedure that does not yet exist, and so the worry I have raised is merely theoretical. I doubt that mentalists will worry about that (they tend to defend their view using the very same theoretical procedure). Just in case, however, I might note that there is another worry involving a procedure that is not merely theoretical: abortion. If animals have interests that are on a par with those of the persimals they host, then one of the standard ways of defending abortion fails (I had better add that ruling out one does not rule out all). That approach goes as follows:

> Although killing persimals is of great significance, and difficult to justify morally, abortions do not kill persimals; abortions only preclude their existence, which is of no moral consequence whatever. What an abortion kills is an animal, but this is not of great moral significance, since killing animals is harmless to them, or virtually so, given the low level of welfare animals are able to attain if not killed.

This line of thought fails utterly if abortion kills the animal (while it is a fetus) and if the interests of human animals are on a par with the interests of their persimals, since killing the animal is as bad for that animal as killing the persimal would be for that persimal (Quinn 1984; Marquis 1989). It is true that aborting the development of the animal while it is a fetus forestalls its developing the apparatus of responsiveness, but this is quite different from forestalling its coming into existence at all. Forestalling the very existence of an animal cannot harm that animal. By contrast, what stops an existing animal from developing responsiveness deprives it of any good life it otherwise would have had. So does killing it just after it becomes responsive. Given comparativism, such killings are not harmless.

I have been arguing that an absurdity follows from the claims that (a) we persimals are distinct from our animal hosts, and that (b) we and our hosts both have interests. The absurdity is that these interests can come apart, even clash, as in cases involving a brain transplant. (If you prefer to say that the absurdity is really just that—or also that—we and our hosts both *have* interests, be my guest.) If the argument has succeeded, then we are in a position to conclude that if we persimals are distinct from our hosts, then one or the other of us lacks interests. One or the other must be unresponsive.

5.6. OTHER ANIMALS

There are, of course, many kinds of animals other than the human being. At the beginning of the paper I noted that there are strong grounds for the view that many of these other animals are responsive. If human animals do not have interests, should we reconsider that view?

I would think so, for *whatever* features we are likely to consider sufficient for an animal's responsiveness we *find in the human animal*. If *any* animal is responsive, then human animals are.

5.7. THE ZOMBIE ARGUMENT

So all animals are zombies! In any case, that is what proponents of mentalism will need to conclude, and part of their argument seems uncontroversial. To see why, let's review. In the previous section we said that if any animals are responsive, then human animals are. But in the section just before that, we said that if we persimals are distinct from our animal hosts, then one or the other of us lacks interests, and presumably we are not the ones who are lacking. That is, if persimals are not animals, then no human animals have interests. So if persimals are not animals, then *no animals whatever* have interests. We have arrived at the second premise of the zombie argument:

1. We are not animals.
2. If we are not animals, then no animals have interests.
3. No animals, therefore, have interests.

So if we really aren't animals, we must accept the bizarre conclusion that no responsive animals exist. (If you are willing to deny that animals are responsive then you probably will want to say, along with Sydney Shoemaker, that animals have *no* sort of mental properties [see, e.g., Shoemaker 1984; for an excellent critique, see Olson 2002]. They really are zombies. You might go on to say that some animals *appear* to have a mental life because they have a component that *does* have a mental life, namely a persimal. I discuss this latter idea in Luper 2014b.)

5.8. THE ARGUMENT INVERTED

Since a great many animals *seem* to be responsive to various degrees, mentalists will need to explain away the appearances. We must follow them down that road—unless we turn their argument on its head, as follows:

1. Some animals have interests.
2. If we are not animals, then no animals have interests.
3. We are, therefore, animals.

This inverted argument is at least as strong as the zombie argument. But there is also reason to think that the inverted argument is more plausible.

Consider that mentalism is motivated by our intuitive judgments about cases such as the transplantation of Ishmael's brain. In particular, we are strongly inclined to think that something (me) that would go with Ishmael's brain were it moved to a new body has interests. Before it went with Ishmael's brain, it was hosted by Ishmael, and it had interests, but it was not Ishmael himself, as he did not move to the new body. Before we consider such scenarios, it seems quite evident to us that many animals are responsive. When first confronted with transplantation scenarios, most of us tend to ignore the animals that are involved (as Paul Snowdon points out); our attention diverted, we do not ask whether they have interests of their own. When that question is forcibly brought to our attention, however, we are initially torn. We can see how ridiculous it is to deny that the animals are responsive, but it also seems that something with interests travels with the brain. (Do transplant scenarios suggest that an animal with interests is *impossible*? Imagine an animal whose responsiveness apparatus is distributed throughout the entire animal, so that the apparatus cannot possibly be moved without taking the whole thing along for the ride.)

Let's reexamine this last thought. If we say we are animals, can we explain away the impression that something responsive travels with the brain? I doubt that we can satisfy diehard mentalists. But I do think that we can hold our ground. Part of the story we can tell is that what travels is not me, or anything else with interests, but rather the apparatus that once enabled me to desire, to experience pleasure, to recall these, and to be conscious of the results. With my teeth, I can chew; moved into the mouths of toothless people, these teeth would enable *them* to chew. Something similar is true of the apparatus that makes desire, pleasure, and the rest possible. Imagine that the transplant of this apparatus is done piecemeal, starting with Ishmael's desire module. Clearly nothing with interests would go with this module. It is simply an apparatus that makes desiring possible for some creature while it is in place. Swapped with yours, what was once my desire module would then enable you to want things, and yours would enable me to desire things. (Imagine that you and I had to take turns with this module, like the three Graeae sisters who shared a single eye and tooth.)

Similarly, if our memory modules were swapped, my former module would enable you to form fresh memories, and yours would do the same for me. If the contents of these modules, our old memories and desires, came along for the ride, we would both suddenly desire things we never wanted before, and we would experience very convincing illusions of having done things we never did. Finally, if our consciousness modules were exchanged, what was once my module would then enable you to be conscious, while yours would enable me to be conscious. None of this would change if all three modules were swapped at the same time. You and I stay put. (Similarly, in Parfit-style division [1984], where each half of my brain goes into a separate body, as well as the variant case in which only half is moved and the other stays put, nothing with interests goes along for the ride. Even those who are tempted to think that an entire brain sports something with interests will deny that each half of an intact, normally functioning brain has its own interests.)

However, if you knew about the swap ahead of time, and what it entailed, you could anticipate that I would end up with the urge to carry out your cherished projects, and that you would want to carry out mine. Looking ahead, and imagining what it will be like to be me after the swap is completed, you might well care more about my continuing life than about your own. When you anticipate the perspective I will have in the future, what you are anticipating will be very much like the perspective you yourself would have had if not for the transplant. No wonder it is tempting to think that something that has your interests—you—will end up in my brain.

ACKNOWLEDGMENT

I thank Tatjana Višak and Steven Campbell for their comments on earlier drafts.

6

The Value of Coming into Existence

NILS HOLTUG

6.1. INTRODUCTION

Death, it is generally agreed, harms an individual, at least insofar as this individual would otherwise have had a future life that was worth living. Indeed, on a simple account of the badness of death, death is bad because and to the extent it deprives an individual of future goods that would otherwise have accrued to her. This is the *deprivation,* or *future goods,* account of the badness of death (Feldman 1992; McMahan 1998).[1] On this simple account, an individual's death is likely to be worst for her if it occurs just after she comes into existence, because it is likely that this is when she stands to lose the most future goods (that is, when the net value of her future is highest). So when I look back on my life, I may greatly appreciate that I did not die just after having come into existence, or indeed at any other time between then and now.[2] But might I not also appreciate that something else did not happen—something that would have taken away at least as many of my future goods—namely, my never coming into existence in the first place?

In this paper, I shall defend the *value of existence view,* which is the view that it can benefit (or harm) an individual to come into existence. More specifically, I shall defend the *comparative value of existence view,* according to which existence can be better (or worse) for an individual than never existing.[3] To clarify, I do not claim that an individual's *mere* existence can be better (or worse) for her than never existing—indeed, this would be an odd claim to make. Rather, it is the particular life led by the individual that may be better or worse. First, I provide an argument for the comparative value of existence view. Then I consider the objection that existence cannot be better (or worse) than never existing, because that would imply that never existing could be worse (or better), the objection being that nothing can be worse (or better) for an individual who never exists. More precisely, I consider and reject various versions of this objection. Finally, I consider the implications of the comparative value of existence view for *person-affecting morality*, as well as the conjunction of these two views for our obligations to nonhuman animals.

Since the argument is basically the same regardless of whether we are considering the existence of human or nonhuman animals, I shall simply refer to the possible beneficiaries of existence as "individuals"—a term that may refer to both kinds of beings. I shall, however, only be concerned with individuals for whom life may go better or worse, and so to whom welfare may accrue. This is because only such individuals may benefit from existence or, for that matter, from anything else. Arguably, welfare requires a certain level of psychological complexity, but it does not require more complexity than may be found in, say, a dog or a squirrel.

What is the moral relevance of the value of existence view? First, there is the question of whether it somehow commits us to the view that we have a pro tanto moral reason to bring (happy) individuals into existence, and perhaps in some cases even an obligation to do so. While this is an implication regarding our obligations to possible people that has troubled many population ethicists, presumably it is no less troubling insofar as it applies to nonhuman animals.

Second, the value of existence view allows us to give some seemingly plausible person-affecting moral accounts of certain cases that involve bringing individuals into existence that would not otherwise be available to us (Holtug 2004; 2010). Suppose we have knowingly created an individual who is destined to suffer greatly throughout his short and miserable existence, and thus to have a life worth not living. We could give an impersonal explanation of why it was wrong to create this individual, for example by pointing out that he detracts from the net sum of welfare in the universe. However, this explanation would not intuitively capture the sense in which we may believe that what we did was *bad for him*. If, on the other hand, we claim that it may harm an individual to come into existence, we can tie the wrongness of bringing him into existence to the fact that we *harmed* him by doing so.

Third, suppose we hold:

The narrow person-affecting principle. An outcome, O_1, cannot be better (or worse) than another outcome, O_2, if there is no one for whom O_1 is better (or worse) than O_2.

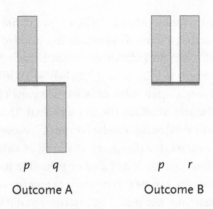

p   q   p   r

Outcome A   Outcome B

Figure 6.1 Two Possible Future Populations

And suppose we reject the comparative value of existence view. Now compare an outcome, A, in which all future individuals (the q-individuals) have miserable lives (lives worth not living) and an outcome, B, in which an equal number of different future individuals (the r-individuals) have very valuable lives, and where those of us who already exist (the p-individuals) are equally well off in A and B. These outcomes are represented in figure 6.1 (where the width of a column represents the number of individuals in a group, the height represents their level of welfare, and the line connecting the columns in A represents the level where life ceases to be worth living).

We shall here have to conclude that A is no worse than B. After all, on the assumption that A cannot be worse for the miserable future individuals who inhabit it, the narrow person-affecting principle implies that it cannot be worse, since it is not worse for anyone who exists in it. It seems, then, that there are several reasons why we need to consider the plausibility of the value of existence view.

6.2. THE COMPARATIVE VALUE OF EXISTENCE VIEW

The comparative value of existence view has received a great deal of attention in recent years. It has been defended by Arrhenius and Rabinowicz (2015), Bradley (2013), Hare (1993), Holtug (2001; 2010, chap. 5), Johansson (2010), Persson (1997) and Roberts (2003). Critics, on the other hand, include Broome (1993), Buchanan et al. (2000), Bykvist (2007), Dasgupta (1995), Heyd (1992), McMahan (2013), Narveson (1967), Parfit (1984) and Višak (2013). Some critics of the comparative value of existence view, however, are willing to accept the more general value of existence view. They argue that while it cannot be better (or worse) to exist than never to exist, it can be *good* to come into existence (McMahan 2013; Parfit 1984).[4] Furthermore, some philosophers accept a limited version of the value of existence view in that they claim that while an individual cannot benefit from coming into existence, she can be harmed (Benatar 2006; Fehige 1998).

I shall take the comparative value of existence view to be a view about the relative value of states of affairs for an individual. That is, it claims that the state that an individual exists can be better (or worse) for her than the state that she never exists. Furthermore, the kind of value it is concerned with is intrinsic value; that is, final value or value for its own sake. Thus, the claim is that existence can be intrinsically better (or worse) for an individual than never existing, not just, say, instrumentally better. The values being compared are (1) the intrinsic value, for an individual, of the state that she exists (or rather, leads a particular life) and (2) the intrinsic value, for that individual, of the state that she never exists.

The particular kind of intrinsic value with which the comparative value of existence is concerned is welfare (well-being). It compares the welfare value of existing and never existing for an individual, rather than, say, the moral or aesthetic value of these states. This also means that the exact implications of the comparative value of existence view will depend on what kind of theory of welfare we are assuming. Nevertheless, I want here to be neutral between different theories of welfare. But let me very briefly say a little about how to compare the value of existing and the value of never existing for an individual on the basis of three different such theories.

Some philosophers rely on a *preference satisfaction theory* in their defense of the comparative value of existence view (Hare 1993; Persson 1997). In the simplest version of this theory, the state that he exists is better for an individual than the state that he never exists, insofar as he prefers the former state to the latter (and the former state is worse for him if he in fact prefers the latter state). Alternatively, the value of these two states for an individual may be based on the preference satisfactions contained in each. Insofar as the state that he exists holds a net positive value in terms of preference satisfactions, it is better for him than the state that he never exists, because the state that he never exists, of course, does not hold any positive value. If the state that he exists holds a net negative value in terms of preference satisfaction (frustration), the state that he exists will be worse for him.

In hedonism and an objective list account, the state that an individual exists is better than the state that he never exists for him, insofar as the former state holds a net positive value in terms of, respectively, mental states and items on the objective list. Insofar the former state holds a net negative value in terms of mental states or items on the objective list, it will be worse for him than the state of never existing.

An argument for an individual's existence being better (or worse) for him than never existing can thus be made irrespective of which of these theories of welfare we assume, although the relative value of these two states for him may in some cases differ, depending on which theory we hold.

Note that while most nonhuman animals may be incapable of preferring existence to never existing (because this particular preference requires rather advanced cognitive skills), it is possible to compare the value of these two states for them in terms of the preference satisfactions, hedonic states, and items on the objective list contained in each. And while it may be objected that, regarding the latter claim, nonhuman animals cannot instantiate items on the objective list, presumably many (mammals, birds, and fish) can instantiate at least some such items, including pleasurable mental states (and the avoidance of pain and other negative mental states), which I take it would be items on any plausible version of an objective list (Griffin 1986).

6.3. THE LOGIC AND METAPHYSICS OF THE BETTERNESS RELATION

According to the comparative value of existence view, the state of existing can be better (or worse) for an individual than the state of never existing. However, it has been objected that this is an incoherent view. Here are representative statements by Parfit and Broome, respectively:

> Causing someone to exist is a special case because the alternative would not have been worse for this person. We may admit that, for this reason, causing someone to exist cannot be *better* for this person. (Parfit 1984, 489)

> At least, it cannot ever be true that it is better for a person that she lives than that she should never have lived at all. If it were better for a person that she lives than that she should never have lived at all, then if she had never lived at all, that would have been worse for her than if she had lived. But if she had never lived at all, there would have been no her for it to be worse for, so it could not have been worse for her. (Broome 1993, 77)

In fact, this is a popular argument (further proponents include Buchanan et al. [2000]; Dasgupta [1995]; Heyd [1992]; McMahan [2013]; and Narveson [1967]). The argument relies on two premises. The first premise concerns the logic of the betterness relation and may be stated as follows:

(1) y is worse for S than x, if and only if x is better for S than y.

Thus, if existence is better (or worse) for an individual than never existing, never existing is worse (or better) for him. According to the second premise:

The 'no properties of the non-existent' principle. An individual cannot have any properties or stand in any relations in world W if the individual does not exist in W.[5]

It is because a non-existent individual cannot have any properties or stand in any relations that never existing cannot be worse (or better) for him than existing. However, I now want to argue that the comparative value of existence view is in fact compatible with both of these premises.

What would make it true that the state of existing is better for an individual, *s*, than the state of never existing? In one version, this would be the obtaining of the triadic relation "x is better for S than y" between the state that *s exists* (or rather, has such and such a life), *s*, and the state that *s never exists*. In another, less ontologically ambitious version, "x is better for S than y" would simply be interpreted as a three-place predicate. However, not to be accused of making too modest metaphysical assumptions here, I shall elaborate on the former version.

In order for the relation to hold, the three relata specified above must exist. Suppose that *s* is in fact an existing individual. Then the first relata, the state that *s exists*, also exists (in fact it obtains). And as stated, the second relata, *s*, exists. What about the third relata, the state that *s never exists*? Clearly this relata does not obtain. However, arguably, a state need not obtain in order to be an object in a betterness relation. If it did, for example, the following relation could not hold: the state that *the allies win the war* is better than the state that *the Nazis win the war*. Nevertheless, since states are abstract (propositional) entities, it is plausible to claim that they may exist even if they do not obtain. This enables us to claim that the state that *s never exists* is an existing state. And so to claim that the relation "x is better for S than y" may obtain between the state that *s exists*, *s*, and the state that *s never exists*.[6]

Furthermore, "x is worse for S than y" may hold between the state that *s never exists*, *s*, and the state that *s exists*. After all, we have just established that all three relata exist. And so just as existence may be better (or worse) for an individual than never existing, never existing may be worse (or better) for her than existing. Indeed, there is no violation of the no properties of the non-existent principle here, since no properties or relations are ascribed to anything in a world in which it does not exist. All three relata exist in the world in which the relation is claimed to hold.

However, perhaps the objection raised by Broome and Parfit is not that never existing cannot be worse (or better) for an individual than existing, but that *if* the individual does not come into existence, this cannot be worse for him. After all, Broome does state in the quoted passage that "if it were better for a person that she lives than that she should never have lived at all, then *if she had never lived at all*, that would have been worse for her than if she had lived" (my emphasis). However, this does not follow. Broome would have to appeal to a stronger principle; namely:

(2) If x is better (worse) for S than y, then x is better (worse) for S than y even if x obtains.

This premise implies that if never existing is worse for an individual than existing, then never existing is worse for him even in a world in which he does not exist. And this violates the 'no properties of the non-existent' principle. But why should we accept premise 2? Krister Bykvist argues:

To be benefited (harmed) comparatively is not just to be in a state that is better (worse) for you than some alternative state; it must also be true that things *would* have been worse (better) for you in the alternative state. It is in this sense a comparative benefit constitutes a gain in value and a comparative harm a loss. Since nothing has value for you if you do not exist, not even neutral value, you are not better off (or worse off) created than you would have been had you not been created. (2007, 348)

However, (2) cannot be justified on the basis of the logic of the betterness relation. And, indeed, we have a good reason to reject it. On the plausible assumption that s is a concrete particular, the standard view would be that s does not exist in a world in which he is never born. That is, unlike states of affairs, which are abstract entities, s is a concrete particular and so can only exist in a world in which he in fact lives. This means that, had s never been born, there would be no him for this to be worse for, and so it could not be worse for him. This follows from the 'no properties of the non-existent' principle. In other words, the metaphysical basis for claiming that never existing is worse for s than existing has not been preserved in the move from the world in which s exists to the world in which he does not. This is why, in the latter world, the relation does not hold.[7]

It may instead be suggested that even if it is not incoherent to claim that existing is better for s than never existing, this nevertheless does not refer to a genuine *benefit* for s—that would require that it were worse for s if he had never existed. However, while it is plausible to claim that genuine benefits and harms are tied to gains and loses, it is less clear why these would need to involve trans-world comparisons of states. I return to this point below, but note that to make sense of claims about gains and losses, we need only compare the state that *s exists* and the state that *s never exists*. Compared to the latter state, the former state constitutes a gain for *s*.

A related objection to the comparative value of existence view comes from David Heyd:

> First, there is no way to compare the amount of suffering of states of actual people and the state of non-existence of these people. We should resist the temptation of assigning a zero-value to non-existence, thus making it quantitatively commensurable with either the positive or the negative net value of the lives of actual people. (1992, 113; cf. Dasgupta 1995, 383)

However, on the basis of the metaphysical claims I have relied on above, we may in fact ascribe zero-value to the state that *s never exists*, as long as we specify that this is the value this state has for *s* in the world in which he exists.[8] Thus, the relation "x has zero-value for y" obtains between the state that *s never exists* and *s*, where both of these relata exist. Again, there is no violation of the no properties of the non-existent principle. In the world where *s* never exists, on the other hand, this relation does not hold, because *s* does not exist there. But since I am only claiming that existence can be better for *s* than never existing in the world in which *s* exists, the fact that never existing does not have zero-value for him in a world in which he never exists is no objection.[9]

In their defense of the comparative value of existence view, Arrhenius and Rabinowicz (2015) point to a further complication for this view. I have so far assumed that to stand in a relation—and more specifically to be a value-bearer—a state need not obtain, it merely has to exist. However, this assumption may be questioned. Consider, for example, the state that *s is happy*. Arguably, if this state exists but does not obtain, it neither benefits *s* nor adds to the value of the world. In order to do so, *s* must *in fact* be happy. More generally, it may be argued, we should accept a claim that is in fact much stronger than (2) namely:

(3) If x is better (worse) for S than y, then x and y obtain.

The problem for the comparative value of existence view is then that when we claim that the state that *s exists* is better for *s* than the state that *s never exists*, the latter state does not obtain. Therefore, this claim violates (3).

However, whatever we say about the exact nature of value-bearers, the claim that the betterness relation holds only between states that obtain seems too strong (cf. Arrhenius and Rabinowicz 2015). On the modal actualism I have assumed so far, relations can hold only between coexisting objects, or objects that inhabit one and the same world. But it seems evident that betterness relations sometimes hold between states that do not—indeed cannot—obtain in one and the same world. Consider, for example, the state that *s is (all things considered) happy at t* and the state that *s is (all things considered) unhappy at t*. If we required both states to obtain in the same world in order for the former state to be better (or better for *s*) than the latter, then it could not possibly be better. Likewise, the state that *s exists* and the state that *s never exists* cannot obtain in the same world, and requiring that they do so in order for the former state to be better for *s* than the latter seems too strong a requirement. So while, plausibly, coexisting in a world is a condition for the betterness relation to obtain, it seems doubtful that co-obtaining in a world is such a condition.

6.4. PERSON-AFFECTING MORALITY

As pointed out in the introduction, the value of existence view allows us to give a plausible explanation of why it is wrong, everything else being equal, to create miserable individuals. We *harm* them by doing so. Intuitively, this may seem a more plausible explanation than the impersonal explanation we would otherwise have to give. This explanation applies not only to "wrongful life" cases, where children are brought into existence in a condition that is judged to be worse for them than never existing. It also applies to the creation of nonhuman animals whose lives are judged to be worth not living, perhaps because in their creation they have been genetically engineered (or otherwise modified) to have a certain disease that causes significant suffering (as in certain animal experiments), or simply because the life to which they have been created is of a very low quality (which may potentially involve factory farming, animal experiments, zoos, etc.). Creating nonhuman animals to such lives is wrong, everything else being equal, because it harms them.

Person-affecting principles may take either a narrow or a wide form (Holtug 2004; 2010; Parfit 1984). Consider first the narrow person-affecting principle, referred to above, according to which an outcome, O_1, cannot be better (or worse) than another outcome, O_2, if there is no one for whom O_1 is better (or worse) than O_2. Since, as I have argued, coming into existence can be better or worse for an individual than never existing, this principle is compatible with claiming that it is worse if we cause unhappy individuals to exist than if we cause (another group of) happy individuals to exist, where our own happiness is not affected by this choice. Thus, it is compatible with claiming that, in figure 6.1, A is worse than B, since A is worse for the q-individuals. However, now compare the following two outcomes, the comparison of which is a version of the non-identity problem (Parfit 1984). In C, we cause a group of very happy individuals to exist. In D, we cause an equal number of different individuals to exist who are happy, although significantly less happy than the individuals in the former group. No one else's welfare is affected. While it is plausible to claim that D is worse than C, this judgment is ruled out by the narrow person-affecting principle. After all, there is no one in D for whom this outcome is worse.

Now consider:

The wide person-affecting principle. An outcome, O_1, cannot be better (or worse) than another outcome, O_2, if, were O_1 to obtain, there would be no one for whom O_1 was better (or worse) than O_2 and, were O_2 to obtain, there would be no one for whom O_2 was worse (or better) than O_1.

Unlike the narrow person-affecting principle, this principle allows us to deal plausibly with the non-identity problem. It is compatible with the claim that D is worse than C since, if C obtains, this is better for the happy individuals who exist only in C. The wide person-affecting principle, then, provides a more plausible person-affecting moral account.

The wide person-affecting principle also explains how we can avoid a further objection, which is due to Bykvist (2007). Suppose we cause an individual to exist whose life is worth not living. Then, according to the comparative value of existence view, this is worse for her than never existing and presumably, it is wrong to cause her to exist (everything else being equal). But suppose instead that we do not cause her to exist. Then, according to the argument presented above, her existence is no longer worse for her than never existing and, by assumption, not worse for anyone else either. But if her existence is not worse for her, then why would it have been wrong to bring her into existence? The problem is that whether it is wrong to cause her to exist seems to depend on whether or not we do cause her to exist. More generally, the following principle is violated:

Normative invariance. An action's normative status does not depend on whether or not it is performed.

However, when comparing two outcomes, the wide person-affecting principle requires us to look at what would be the case if either of these outcomes obtained and, in particular, whether anyone would be better or worse off in them. Here, our judgment about the moral value of these outcomes does not depend on which outcome in fact obtains, and therefore neither should our judgments about the wrongness of bringing individuals into existence. Everything else being equal, it is wrong to bring a miserable individual into existence because, if we do so, this will be worse for her.

What are the implications of the wide person-affecting principle for animal ethics? Consider first the following version of the non-identity problem. Either we continue to produce a certain kind of chicken that grows fast, but where, because of their growth rate, this causes the chickens some amount of pain in their legs. And suppose, for the sake of argument, that these chickens have lives worth living, although their welfare is of course reduced. Or we produce a different kind of chicken that grows slower, so that the chickens do not experience this pain. Clearly, everything else being equal, it is worse to produce the former chickens. However, the narrow person-affecting principle rules out this judgment. Since the fast-growing chickens have lives worth living, there is no one for whom it is worse to produce them (I am assuming that others, including consumers, are indifferent). According to the wide person-affecting principle, on the other hand, we may plausibly claim that it is worse to produce the fast-growing chickens. After all, if the happier chickens were to come into existence instead, this would be better for them.

There is nevertheless a certain worry that we may have about the conjunction of the comparative value of existence view and the wide person-affecting principle. Often, philosophers invoke person-affecting principles because they want to resist the conclusion that we have reason to cause happy individuals to exist (Heyd 1992; Narveson 1967). The conjunction of the comparative value of existence view and the wide person-affecting principle, however, does not have this implication. Compare outcome E, in which we cause a group of happy individuals to exist, and outcome F, in which we abstain from causing them to exist, where no one else's welfare is affected. While the narrow person-affecting principle at least implies that F cannot be worse than E, the wide person-affecting principle does not even do that. This is because if we cause these extra individuals to exist, this will be better for them.

Now, neither the comparative value of existence view nor the wide person-affecting principle (nor the conjunction thereof) *implies* that it is morally good to create happy individuals, or indeed that we have any moral reason to do so. However, when making the case for the wide person-affecting principle, I relied on the claim that in the non-identity problem, it is worse to cause the less happy individuals to exist, because if we cause the happier individuals to exist instead, this will be better for them. And that involved attaching at least some positive moral weight to their existence.

In population ethics, the claim that we have reason to bring into existence happy individuals has been associated with a number of disturbing claims, not least:

> *The repugnant conclusion.* A world populated by individuals, every one of whom
> has a life barely worth living, would be better than a world populated by, for
> example, ten billion individuals, all of whom have very worthwhile lives—as
> long as the former population is sufficiently large. (Parfit 1984, chap. 17)

This conclusion is generally considered very counterintuitive—hence its name—when the individuals referred to are taken to be humans, but presumably it is also rather counterintuitive when applied to nonhuman animals. Compare, for example, an outcome in which farm animals have sufficient space and an environment in which they can engage in their natural behavior and an outcome in which a much larger number of animals are crammed together to achieve a more efficient production. Indeed, it does not seem at all unrealistic that there is a quality versus numbers trade-off in contemporary farming and that, presumably, the animals in the former outcome would have much better lives. However, care is taken to leave the animals in the latter outcome at a level where their lives are just barely worth living. Intuitively, it does not seem plausible that, everything else being equal, the outcome in which the larger population of animals has lives barely worth living is better.

6.5. JUSTICE FOR ANIMALS

Perhaps the most controversial aspect of attaching moral weight not only to the welfare of nonhuman animals, but also to that of possible nonhuman animals, is that this will give rise to (further) conflicts of interest between human and nonhuman animals. Thus, distributive principles such as, to some extent, utilitarianism, but in particular egalitarianism, prioritarianism, and sufficientarianism are already quite demanding as regards accommodating the interests of nonhuman animals (Holtug 2007, 2010; Vallentyne 2006), and including possible future nonhuman animals within the scope of such distributive principles will only increase interspecies conflicts of interest. In the case of the latter three principles, it will mean that not only should we assign greater weight to the interests of nonhuman animals who are worse off than we are (e.g. because they have shorter lives) than to ourselves, but also to the interests of possible future nonhuman animals who will, after all, enjoy no welfare if they do not come into existence.

Now, there are two obvious ways in which we may attempt to lessen the weight of possible future nonhuman animals. First, we might assign less weight to (most) *nonhuman animals*, either by simply assigning less weight to the welfare of nonhuman animals, or less weight to the welfare of individuals who do not possess a particular advanced psychological characteristic, such as self-consciousness, or less weight to individuals who have fewer psychological potentials or a lower moral standing. However, the problems with these suggestions are many and well-rehearsed in the literature (for my own contribution, see Holtug 2007, 2010). Second, we may lessen the weight we assign to *possible* individuals, say, by opting for a stronger person-affecting requirement than the wide person-affecting principle. Thus we may claim that, for example, only the welfare of actual, necessary or some other modal category of individuals falls under the scope of our distributive principle. Again, however, the problems are many and well-rehearsed (Holtug 2010, chap. 8).

The problems in population ethics seem to me to be the most difficult problems in moral philosophy. And while we do not need to consider nonhuman animals to realize this, including such animals in our analysis certainly does not make the task of coming up with a satisfactory theory any easier. In fact, in response to the difficulties in population ethics, some philosophers entertain the thought that these problems are so troubling that they may provide an independent reason to reject moral realism altogether (Arrhenius forthcoming; McMahan 2013). However, it is my contention that insofar as we do try to come up with a moral theory that applies to the sphere of population ethics, the comparative value of existence view and the wide person-affecting view are likely to be part of the best available such theory, where these views apply to human and nonhuman animals alike.

ACKNOWLEDGMENT

I would like to thank Tatjana Višak for comments on an earlier version of this paper.

NOTES

1. In fact, McMahan considers this a partial account of the badness of death, at best. He believes that the badness of death depends not only on future goods, but also on various other features, including the extent to which an individual, at the time of her death, is related—in the sense that prudentially matters in survival—to herself at the times these future goods would have accrued to her. Incidentally, while McMahan (2002, 165–74) and I (Holtug 2010, 103–11) roughly agree on this more complex account of the badness of death, we disagree about its implications for the wrongness of killing. I critically discuss McMahan's view in Holtug 2010, 330–34, and more elaborately in Holtug 2011, and defend my own view most elaborately in Holtug 2010, 330–34.

2. There is of course an issue here of when I came into existence, and the answer to that question will depend on which particular theory of personal identity we adopt (see Holtug 2010, chap. 2). However, whenever that was, I may greatly appreciate that I did not die immediately after.

3. For a more detailed defense, see Holtug 2010, chap. 5.

4. I critically discuss Parfit's noncomparative version of the value of existence view in Holtug 2010, 146–47.

5. This principle is also sometimes referred to as "actualism" (Bykvist 2007, 339; Johansson 2010, 286).

6. Wlodek Rabinowicz first suggested this account of the relation to me in personal communication and I have defended it in Holtug 2001 and Holtug 2010, chap. 10. Since then, he and Gustaf Arrhenius have defended it in print in Arrhenius and Rabinowicz 2015.

7. Melinda Roberts (2003, 168–69), in contrast, suggests that it could have been better for an individual never to have existed; that is, she suggests that the relevant relation would hold even in a world in which the individual does not exist.

8. In fact, it is less obvious that we need to be able to ascribe zero-value to the state that *s never exists* if we make the value judgment simply on the basis of *s* preferring existing to never existing. But on the other welfare theories mentioned above, the comparison is a two-step procedure in which the value of each of the two states is first assessed and then compared, and so we may need to be able to assign a value, most plausibly zero-value, to the state that *s never exists*.

9. Incidentally, while they do not explicitly reject the claim that the state that *s never exists* has zero-value for him, Arrhenius and Rabinowicz (2015) argue that we need not rely on this claim when defending the comparative value of existence view. Johansson (2010, 294), on the other hand, relies on a version of this claim in his defense, namely on the claim that in the world in which *s* exists, the world in which he never exists has zero-value for him.

PART II

Moral Evaluation of Killing Animals

7

Do Utilitarians Need to Accept the Replaceability Argument?

TATJANA VIŠAK

7.1. INTRODUCTION

It is no surprise that the utilitarian stance on how to treat animals is iden-tified with Peter Singer's take on the issue. After all, Singer is one of the most famous contemporary philosophers, and in 2013 he was named one of the three most influential people in the world by the Gottlieb Duttweiler Institute (Frick, Gloor, and Gürtler 2013). He is the best-known contem-porary utilitarian, and animal ethics is one of the fields of application with which he is primarily associated. His book *Animal Liberation* (1975, 1995) is a founding source of the animal liberation movement, and it inspired a vast number of readers all over the world to reconsider their attitudes towards animals. Also in his widely read book *Practical Ethics* (1979, 1993, 2011), as well as in many other publications, Singer defends what he takes to be our duties towards animals from a utilitarian perspec-tive. He prominently rejects speciesism, which he defines—analogous to racism and sexism—as the unjustified discounting of the interests of animals. In line with utilitarianism's founding fathers—Jeremy Bentham (1748–1832), John Stuart Mill (1806–1873) and Henry Sidgwick (1838–1900)—Singer makes clear that in considerations about inflicting pain to an individual, it does not matter whether the individual can reason or talk, but only whether it can suffer. Singer accords all sentient beings equal moral consideration.

For Singer, and for utilitarians in general, equal *consideration* of interests does not entail any limitations as to how an individual may be *treated*. The moral aim, according to utilitarianism, is to maximize overall welfare, and in some situations harming or killing an individual may be the action that achieves that aim. So, for instance, if killing a human or an animal is the only way of preventing a greater disaster for this individual or for others, utilitarianism requires the killing.

A remarkable aspect of Singer's view is that it allows the killing of a happy and innocent human or nonhuman animal, even if this is *not* the only way of preventing a greater disaster, and indeed even if it does not bring about more welfare at all. The replaceability argument offers a way of compensating the welfare loss that results from killing an animal that would otherwise have had a pleasant future. Killing an animal that would otherwise have had a pleasant future is morally neutral, if the killing allows another animal to exist—an animal that would not otherwise have existed, and whose lifetime welfare is as great as the future welfare of the killed animal would have been. In that case, the animal that takes the killed animal's place compensates for the lost welfare. It is assumed that the killing does not have any (uncompensated) negative side effects. Under these conditions, the killing does not reduce the total quantity of welfare in the world. If one's options were either to kill and replace the animal or to let it live, both options would be equally permissible.

The replaceability argument is the most controversial aspect of Singer's theory (Singer 1993, 386; Singer 2011, 107). In his writings over the past forty years, Singer's uneasiness with this argument—and in particular his various efforts to limit its scope to nonpersons (i.e. to human and nonhuman animals that lack rationality and self-awareness)—were responsible for the most remarkable revisions of his theory. These include his tentative acceptance and then rejection of prior existence utilitarianism for self-conscious beings (Singer 1993), his claim that welfare can never be positive (Singer 1993, 2011), and his acceptance of preference-independent value (Singer 2011, 117). While it is controversial whether any of these revisions achieves the desired aim (Kagan this volume, Višak 2013), Singer's most recent turn to hedonist utilitarianism makes him squarely face the replaceability argument again (Lazari-Radek and Singer 2014).

Singer's acceptance of the replaceability argument leaves him no direct way of condemning the routine killing of animals in practices such as meat production, milk production, egg production, aquaculture, sports hunting, sports fishing, and animal experimentation, not to mention possible practices of routinely killing humans.[1] Those who accept the replaceability argument can, as Singer does, still condemn many of the above-mentioned practices in which animals are routinely killed and replaced. After all, these practices usually do not satisfy the conditions of the replaceability argument. For instance, many animals that are currently kept for consumption do not lead pleasant lives at all. Furthermore, animal agriculture pollutes the environment and wastes resources, and it thus seems, all things considered, that it is not a practice that promotes welfare (Matheny and Chan 2005). Nevertheless, the replaceability argument, which Singer fully accepts, implies that *under the specified conditions* there is nothing wrong about the routine killing of animals, including humans.[2]

The aim of this paper is to show that utilitarianism need not entail the replaceability argument. Contrary to what has become the standard utilitarian view, this moral theory can grant animals a stronger protection against killing. In the next section, I introduce an important value-theoretical assumption that I make throughout this paper: existing as opposed to never existing cannot benefit or harm an animal. In section 7.3, I propose an alternative—person-affecting as opposed to impersonal—version of utilitarianism. It includes a proposal concerning how to apply person-affecting utilitarianism to cases in which different individuals exist in different outcomes. In section 7.4 I show that person-affecting utilitarianism avoids the replaceability argument. In section 7.5 I point out that the implications of the proposed theory are more plausible than those of rival theories in a range of challenging cases that have prominently been discussed in the literature. In section 7.6 I address possible challenges to the proposed theory.

7.2. A VALUE-THEORETICAL ASSUMPTION

I assume that existing (i.e. having a life with a certain lifetime welfare level) as opposed to never existing (i.e. having no life at all) cannot be better or worse for an individual. The reason seems simple: according to the standard counterfactual account of harm and benefit, some event or state of affairs benefits me if, due to that event or state of affairs, I will be better off than I would otherwise have been. This assessment requires a comparison of my welfare in the world in which the event occurs with my (or better, my counterpart's) welfare in the closest possible alternative world in which that event had not occurred. For example, my lifetime welfare in the possible world in which I die tonight will be my lifetime welfare up to tonight. My lifetime welfare in the closest possible alternative world in which I do not die tonight will be my lifetime welfare up to the time at which I (or, more precisely, my counterpart) dies in that closest possible alternative world. Granted, it is usually impossible to know when I would otherwise die and how well off I would be until then. But we can assume that these questions have answers. The fact that we often do not know what would have happened in the counterfactual situation does not disqualify this comparative account of benefit and harm. According to the comparative account of benefit and harm, which of two possible courses of action is better for me depends on my lifetime welfare in both possible future worlds. If I compare a possible world in which a particular individual, say Tom, exists, with an alternative possible world in which Tom does not exist, then which world would be better for Tom? In one world Tom exists and has a certain welfare level. In the alternative world, Tom does not exist and therefore has no welfare level. Can we say that one of these possible worlds is better *for Tom*? I take it that we can't.

It is a common mistake to set the welfare level in case of non-existence at the neutral point. Holtug's defense of the value of coming into existence also relies on that move. He says that "we may in fact ascribe zero-value to the state that *s never exists*, as long as we specify that this is the value this state has for *s* in the world in which he exists" (Holtug, this volume). Holtug and others who make this claim acknowledge that an individual experiences neither positive nor negative welfare if he doesn't exist. Holtug even agrees that one cannot attribute any property to non-existing individuals. That is why he emphasizes that non-existence has zero value for an individual *only in a world in which this individual exists*. This move, however, does not help. Holtug claims that a state (i.e. a propositional entity) can have value for an individual insofar as that individual exists, including the state that that individual never exists. But how am I to determine the value of that state? As explained, according to the standard counterfactual account of harm and benefit, I determine the value of a (non-obtaining) state for me by considering how well off I would be *if that state obtained*. I need to compare my welfare in the actual world to my welfare in this non-obtaining world. Therefore, saying that my non-existence has zero value for me in a world in which I exist *implies* that had I not existed my welfare level would have been zero. However, non-existing individuals do not actually have zero welfare. Rather, they have no welfare.

There is an important difference between "no welfare" and "zero welfare." If an entity has zero welfare, his or her welfare can be compared to the welfare of other entities or to the welfare that the entity's counterpart would have in a different possible world. In contrast, if an entity has no welfare, then these comparisons are impossible. For instance, from the fact that the color green is neither hot nor cold, it does not follow that its temperature is zero degrees, and thus that it is colder than the ocean. Even though rocks don't have any attributes that make them score on an IQ test, it does not make sense to claim that a particular rock is not as smart as a particular boy, or that all rocks are equally dim (Herstein 2013).

So, under what condition does an entity Q that lacks certain Φ-determinate properties have zero Φ, and under what condition does it have no Φ? In a recent paper Ori Herstein advanced the following proposal:[3]

> I believe that Q *being able to* or being the sort of entity that *can have* Φ-determinative properties is such a condition. Where Q not only does not have the properties determinative of some level of Φ, but also is *incapable* (in the relevant sense) of having such properties, it seems false to relate any level of Φ to Q, including a zero level. In such cases the appropriate category to describe Q in terms of Φ is that Q has "no Φ" at all. An entity incapable of having any Φ-determinative properties is simply not the sort of entity that can have any level of Φ, including zero. The Φ metric or scale simply does not apply to such an entity. (2013, 142)

This means the possibility or capacity of Q to have Φ-determining properties is a necessary condition for Q to have a zero level of Φ where Q has no Φ-determinative properties. Admittedly, what exactly it means to have certain capacities is vague. However, it requires at least a metaphysical or conceptual possibility. Perhaps the requirement should be stricter and entail some form of practical possibility. In any case, "comparability and measurability in terms of Φ—even zero Φ—is conditioned on being the sort of entity that can—at least metaphysically—have Φ-determinative properties" (Herstein 2013, 142).

A non-existent individual is not the sort of thing that can have a welfare level. In particular, such an individual cannot have a lifetime welfare level, which is central to the notion of benefit or harm. In order for an individual to have a lifetime welfare level, the individual needs at least to be alive at some time, to exist at some time. By definition, this is not the case for non-existent individuals. Therefore it is logically and metaphysically impossible for a non-existent individual to have a (lifetime) welfare level.

The view that I just defended assumes that an outcome A is better for a subject S than another outcome B if, and only if, S has higher welfare in A than in B. Gustaf Arrhenius and Wlodek Rabinowicz (2015) deny that assumption. They argue that ascribing a welfare level to non-existence is not required in order to claim that this state can be better or worse for an individual than existence. They take it that "better for" comparisons are possible without comparisons of welfare levels. How is that supposed to work? Arrhenius and Rabinowicz appeal to what a guardian angel that is solely concerned with what is in the interest of the individual in question would choose for that individual's sake. They find it plausible that the guardian angel would choose no life for the individual rather than a life with negative welfare, and that it would choose a life with positive welfare rather than no life at all. While I can imagine that one can have this intuition, I am not convinced that the appeal to the guardian angel's choice amounts to more than a restatement of that intuition. How can we know what such a guardian angel would really choose *without* an independent theory of what benefits an individual? Furthermore, it can, of course, be granted that a life with a positive welfare level is good for the individual in question, and that a life with a negative welfare level is bad for her. That alone might explain our intuition concerning the guardian angel's choice. However, in terms of the counterfactual account of benefit and harm, this is not enough.

Therefore I assume throughout this paper that existence cannot be better or worse for an individual than non-existence. A longer life with positive welfare is better for an individual than a shorter such life. A life with negative welfare is bad for an individual, and the longer it is, the worse it is. These judgments do not require the ascription of any welfare level, including zero welfare, to the non-existent. Since, however, reasonable people are currently disagreeing about how to judge that issue, one can read what follows as a conditional: If it is true that existing cannot be better or worse for an individual than never existing, what can be built on it?

7.3. SATURATING-COUNTERPART PERSON-AFFECTING UTILITARIANISM

The impersonal view is a way of comparing outcomes in terms of welfare, at least in theory. The value of an outcome is determined by summing up all the welfare that the outcome entails. In order to sum up the welfare that an outcome entails, it is sufficient to focus on the outcome in question. The outcome's intrinsic aspects determine its value (Temkin 1999, 782). When the value of each outcome has been determined in that way, one needs to compare these values. At that point, for instance, one could see how welfare is distributed and which outcome maximizes welfare (Parfit 1984, 387). Hence the impersonal view in general says that welfare is good, period. From a utilitarian perspective, the impersonal view entails that the outcome that contains more welfare is better:

The impersonal view. Outcome *A* is better (worse) than outcome *B* if and only if outcome *A* contains more (less) welfare than outcome *B*.

Consider a couple deciding whether or not to have a child. Imagine that the decision would not influence the couple's welfare, and that it would not influence the welfare of anyone else. In outcome A, the couple has the child. The couple's welfare levels are, say, 10 units for each individual. Adding the 10 units of the child results in a total value of 30 units. Now one can calculate the value of outcome B, in which the couple does not have the child. Adding up the welfare of the couple makes 20 units of welfare for this outcome. This is how the values of outcomes are calculated according to the impersonal view. In a next step, one can compare the values of the outcomes and determine which outcome maximizes value: in this case, outcome A.

There is an alternative to the impersonal view: the person-affecting view, which is typically defined as follows (Arrhenius 2009):

The person-affecting view.

1. If outcome A is better (worse) than B, then A is better (worse) than B for at least one individual.
2. If outcome A is better (worse) than B for someone but worse (better) for no one, then A is better (worse) than B.

Thus, the person-affecting view describes a necessary and sufficient condition for the betterness of an outcome: If both conditions are met, an outcome is necessarily better. The person-affecting view, as described here, does not offer a full account of how to evaluate an outcome, however. After all, an outcome may be better for some but worse for others.

Person-affecting utilitarianism, in contrast to impersonal utilitarianism, does not simply consider the total quantity of welfare that an outcome entails. Instead, it focuses on the extent to which individuals are better or worse off in the outcome. More precisely it assesses aggregate net benefit (Temkin 1993, 2000; Arrhenius 2000). Thus, according to the person-affecting view, the comparison of outcomes is necessary for *determining* the value of an outcome, and not only for *comparing* values of various outcomes once they have been determined. This is because the person-affecting view does not aggregate welfare as such, but rather aggregates harms and benefits. In order to determine whether something constitutes harm or benefit, one needs to compare the welfare of an individual in an outcome with the welfare of that individual's counterpart in the alternative outcome. This is because determining whether something constitutes harm or benefit, on the standard counterfactual interpretation of harm and benefit, requires a comparison (Klocksiem 2012).

So, for instance, in order to know whether taking some medicine benefitted me, I need to compare how I fare after I took the medicine to how I (or rather my counterpart) would have fared in the counterfactual world in which I would not have taken the medicine. On this view, in contrast to the impersonal view, the value of an outcome cannot be determined simply on the basis of its intrinsic features (Holtug 2004).[4]

We are now in a position to see that person-affecting utilitarianism's implications can differ from those of impersonal utilitarianism. Consider, again, the above-mentioned case of the couple that is contemplating whether or not to have a child. Impersonal utilitarianism would require having the child, since this is the option that maximizes the total quantity of welfare in the world. According to person-affecting utilitarianism, both options are permissible. After all, nobody would benefit or be harmed in either option. This holds because we assume (as explained in section 2) that its existence, as opposed to its never existing, does not make the child better off than it would otherwise have been.

Since more complicated cases are possible, where different individuals (and different numbers of them) exist in each outcome, person-affecting utilitarianism needs a somewhat more elaborated account for dealing with these cases. Various proposals have been made as to how person-affecting views can deal with different people cases (Roberts 2009). Many of them have not been spelled out for applications to different number cases, or cases in which not only different individuals exist in both outcomes, but also different numbers of individuals (Wolf 2009; Hare 2007).

Recently, Christopher Meacham (2012) spelled out a workable proposal, based on Lewisian counterpart relations. Since the world will be different depending on what one does, one can conceive of the possible outcomes as different possible worlds (as I already did throughout this paper). Furthermore, it is assumed that every particular individual can only live in one possible world, but that he or she may have a "counterpart" in another possible world.

Meacham spells out how to specify the relevant counterpart relations (i.e. which individuals in different possible worlds are counterparts of each other):

Now consider an agent in a decision situation at time t. And consider a counterpart relation which, for each ordered pair of available worlds (W_i, W_j) $(i \neq j)$, maps individuals in W_i to counterparts in W_j in a way that satisfies the following ... conditions:

1. *One-to-One Function*: No individual in W_i is mapped to more than one individual in W_j, and no individuals in W_i are mapped to the same individual in W_j.

2. *Before-t Match*: Each individual a who exists before t in W_i is mapped to an individual b who exists before t in W_j that is indiscernable-up-to-t with a.

3. *Saturation*: As many individuals in W_i are mapped to individuals in W_j as possible. (2012, 267)

In conjunction, these rules about "mapping" determine for all possible outcomes which individuals in the outcomes are counterparts of each other. This is necessary in order to know which individuals' welfare levels need to be compared in order to determine harms and benefits. In the above-mentioned example of me taking some medicine, my (Tatjana's) welfare in the actual world in which I take the medicine needs to be compared to my counterpart's (Tatjana*'s) welfare in the closest counterfactual world in which Tatjana* does not take the medicine. In this case, the "before-t match" determines that Tatjana and Tatjana* are counterparts: after all, they were indiscernable up to t, which is the point where Tatjana took the medicine and Tatjana* in the counterfactual world did not. Since we are assessing harms and benefits, the mapping relations cover all individuals who can be harmed or who can benefit. Except for these three principles, there are no further restrictions as to who can be mapped to whom. Thus, if there is no before-t match, every individual in one outcome can be mapped to every other individual in another outcome, and it does not matter precisely who is mapped to whom.[5]

Since the third condition is the most distinctive feature of this counterpart relation, Meacham calls it the "saturating counterpart relation." Thus, in order to distinguish the resulting theory from other versions of person-affecting utilitarianism (one of which will be introduced in section 7.5), I will call it "saturating-counterpart person-affecting utilitarianism," or "SCPA utilitarianism."

Is SCPA utilitarianism a coherent version of utilitarianism at all? In other words, can utilitarianism consistently be person-affecting? I think it can be, and indeed *should* be, person-affecting. A widely accepted basis for utilitarianism is that it extends the principle of prudence for one individual to society as a whole. It is considered prudent for me to benefit myself and, by extension, those I care about. The moral point of view, according to utilitarians, requires extending this concern to society as a whole, and even to nonhuman animals and to future generations. Impartiality, or equal consideration, lies at the basis of utilitarianism. It is, in its own way, an account of fairness and equality. While utilitarianism can sanction unequal treatment, it always requires taking everybody's interests equally into account. This, according to utilitarianism, is what we owe to each other, and morality is fundamentally concerned with what we owe to each other. While utilitarianism should be person-affecting in this (wide) sense, it should not be person-affecting in a narrower sense; that is, it should not be concerned with how any particular individual *as a particular individual* fares. After all, utilitarianism is typically impartial and accepts aggregation of welfare, and I do not wish to take issue with that. This possible understanding of utilitarianism as a moral theory fits well with the wide person-affecting view presented here.

7.4. AVOIDING THE REPLACEABILITY ARGUMENT

Person-affecting utilitarianism in general, and SCPA utilitarianism in particular, can avoid the replaceability argument. Given the assumption that existence as opposed to non-existence cannot benefit or harm an individual, person-affecting utilitarians would evaluate the replacement scenario as follows.

In figure 7.1, the horizontal axis represents time, the vertical axis represents the individuals' welfare level. The black and white areas above the horizontal line represent individuals. A is the outcome in which the individual Black is not killed at time t_1. B is the outcome in which Black is killed at t_1 and replaced by the individual White. Killing Black at t_1 harms Black, since it deprives him of future welfare (Bradley, this volume). Bringing White into existence at t_1 does not benefit White, since (as explained in section 7.2) existence as opposed to never existing does not benefit an individual. Therefore the aggregate net benefit is greater in A than in B. Thus, according to person-affecting utilitarianism, A is required and B is forbidden.

Since this case is relatively simple, one does not explicitly need to refer to saturated counterpart relations, but doing so underlines the conclusion. Black's counterpart in B is Black*, because of their before-t match. White's counterpart is non-existent in A. B is worse for Black and better for no one. A is better for Black and worse for no one. So, according to SCPA utilitarianism, A is right and B is wrong.

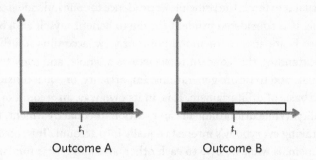

Outcome A   Outcome B

Figure 7.1 The Replacement Case

The replaceability argument can thus be avoided. I consider this a desirable implication of person-affecting utilitarianism in general, and of SCPA utilitarianism in particular.

7.5. SCPA'S COMPARATIVE SUCCESS IN FURTHER DIFFERENT PEOPLE CASES

SCPA utilitarianism avoids the replaceability argument, but at what cost? In order to be a plausible utilitarian alternative to Singer's impersonal utilitarianism, it needs to be shown that it tackles a wide range of cases at least as well as impersonal utilitarianism does. Therefore, I will now confront SCPA utilitarianism with a range of widely discussed different people cases. I will show that it tackles them even better than rival theories do.

For what concerns rival theories, besides impersonal utilitarianism, I will also consider an alternative person-affecting view that Singer considers and rejects. Strictly speaking, it is a view about what entities are morally considerable, and not a view about how to assess outcomes in terms of welfare. However, the view can only coherently be combined with utilitarianism under two assumptions: that existence as compared to non-existence cannot be a benefit or harm, and that morality should be person-affecting (Višak 2013). That is why I discuss this view as an alternative person-affecting view. It is called "prior existence utilitarianism" (Singer 1993, 113) or "necessitarianism" (Arrhenius 2000). It considers only individuals that exist independently of what one does. It thus considers only necessary, as opposed to contingent, individuals. Contrary to SCPA utilitarianism, it lacks any specifications concerning counterpart relations. Framed in terms of counterpart relations, one can say that it matches only individuals that are indiscernable-up-to-t (before-t match). Since that is what SCPA utilitarianism does in the replacement case, prior-existence utilitarianism, being a person-affecting view, reaches the same conclusion in this case. It, too, avoids the replaceability argument. However, its lack of the above-mentioned counterpart specifications causes it trouble in other cases, as we will see. The following examples will point out various implications of these three different views.

First, consider a case in which some population could either remain childless (option A) or have children at the expense of its own welfare (option B), as illustrated in figure 7.2, where the white areas represent populations. The width of the areas represents the population size, the height represents their welfare level. One population, the ORIGINALS, exists in both outcomes, albeit with a different welfare level. (More precisely, the

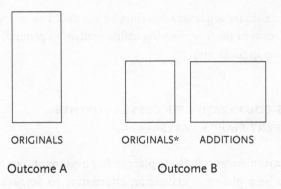

ORIGINALS   ORIGINALS*   ADDITIONS

Outcome A   Outcome B

Figure 7.2 ORIGINALS versus ORIGINALS* and ADDITIONS

ORIGINALS in outcome A have counterparts, the ORIGINALS* in outcome B.) The other population, the ADDITIONS, exists only in outcome B. According to SCPA utilitarianism, option A is right and B is wrong. This is because the original population in A is better off than their counterparts in B and no one benefits in B. According to prior-existence utilitarianism, A is right and B is wrong. This is because those who exist in both outcomes are better off in outcome A. The others (ADDITIONS) are contingent beings and their welfare does not count. In contrast, impersonal utilitarianism judges that B is right and A is wrong. This is because the quantity of welfare is greater in B than in A. I consider this judgment to be less plausible.

Second, consider the options, illustrated in figure 7.3 of either, from a kind of divine position, bringing about a population that consists of people who are reasonably well off, the OK people (option A), or bringing about a completely different population that consists of people who are much better off, the GREAT people (option B). Again, the width of the areas represents the size of the populations and the height represents their welfare level. In this case, there is no "before-t match". Therefore, SCPA utilitarianism simply requires that one individual in A is mapped to one individual in B, and that as many individuals are mapped in that way as possible.

SCPA utilitarianism tells us that B is right and A is wrong. The counterparts of the OKAY people are the GREAT people. The GREAT people in B are better off in B than the OKAY people are in A. Impersonal utilitarianism also yields the conclusion that B is right, because the total quantity of well-being is higher in B. In contrast, prior-existence utilitarianism implies that both outcomes are equally right, since none of them is better or worse for any necessary individual. This is because no particular individual exists in both outcomes (no before-t match). I consider this to be less plausible. I think that, from a divine position (or, more realistically, as currently living people affecting the welfare of future people), we should bring about outcome B.

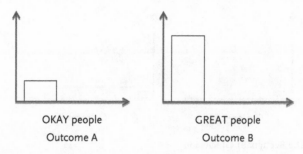

Figure 7.3 OKAY versus GREAT

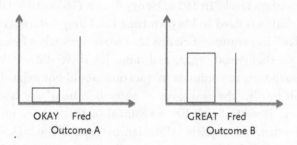

Figure 7.4 Two Populations and Fred

Third, consider a similar case, except that one individual, Fred, exists in both outcomes, as illustrated in figure 7.4.[6] According to prior-existence utilitarianism, which considers only effects on individuals that exist in both outcomes, *A* is right and *B* is wrong. This is because Fred is better off in *A*. Both impersonal utilitarianism and SCPA utilitarianism reach the opposite conclusion, which is more plausible.

Fourth, consider the case known as the "repugnant conclusion" (Parfit 1984, 388), illustrated in figure 7.5. Outcome *A* is a large population consisting of very happy people. Outcome *B* is a much larger population of barely happy people. Nobody exists in both outcomes. Since the total amount of welfare in outcome *B* is higher, due to the large number of people, impersonal utilitarianism requires bringing about outcome *B*. Prior-existence utilitarianism, in contrast, implies that both outcomes are equally permissible, since they contain different individuals. According to SCPA utilitarianism, *A* is right while *B* is wrong. This is because the counterparts of the *A* population are worse off in *B*. The other individuals in *B* are not harmed or benefitted, since the alternative for them would be non-existence. I take this to be the most plausible implication: one should bring about *A* rather than *B*.

Fifth, consider an argument that is used to justify the rearing and killing of nonhuman animals where these animals are granted a life with

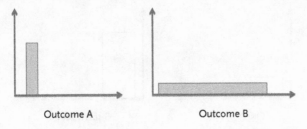

Figure 7.5 The Repugnant Conclusion

a positive welfare level.[7] In 1914, Henry S. Salt (1851–1939) dubbed an argument that was used in his own time (and long before it) the "logic of the larder." The argument claims that we do animals a favor by keeping them for their meat, eggs, and milk, for if we did not keep them for these purposes, the animals in question would not exist. The idea is that, as Salt put it, "the real lover of animals is he whose larder is fullest of them." Despite Salt's fierce rebuttal of this argument, it is still widely defended.[8] An explicit utilitarian defense of the logic of the larder is Richard Hare's "Why I Am Only a Demi-Vegetarian" (1999). Hare defends the consumption of meat from happy animals, because he considers a short and happy life more valuable for the animal than no life at all. The logic of the larder can only be defended on a utilitarian basis if one assumes that existence as opposed to non-existence can benefit an animal, or if one accepts the impersonal view (or both). In fact, it follows from these positions. Given our assumption that existence as opposed to non-existence cannot benefit an animal, impersonal utilitarianism accepts the logic of the larder, while both versions of person-affecting utilitarianism reject it.

Figure 7.6 presents an overview of the three views' plausible (✓) and implausible (X) implications in these five cases and in the replacement case (discussed in section 7.4):

Summing up, SCPA utilitarianism not only avoids the replaceability argument, but it also has more plausible implications in a range of widely discussed different people cases.

7.6. CHALLENGES TO SCPA UTILITARIANISM

So far, so good; but does SCPA utilitarianism face any particular challenges? I can think of two possible criticisms, the first related to transitivity and the second to bringing miserable individuals into existence.

	Impersonal Utilitarianism	Prior-existence Utilitarianism	Saturating-Counterpart Person-affecting Utilitarianism
Replaceability	X	✓	✓
Original versus Original plus Additions	X	✓	✓
OK versus Great	✓	X	✓
Two populations and Fred	✓	X	✓
Repugnant Conclusion	X	X	✓
Logic of the Larder	X	✓	✓

Figure 7.6 Overview of the Three Theories' Plausible (✓) and Implausible (X) Implications

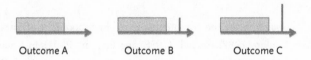

Outcome A   Outcome B   Outcome C

Figure 7.7 The Intransitivity Case

SCPA utilitarianism can be criticized for violating transitivity. Consider the following choice between three outcomes, as depicted in figure 7.7. In each outcome the same individuals exist at the same level of welfare, except for one individual who does not exist in outcome *A*, exists at a positive welfare level equal to the other individuals in outcome *B*, and exists at an even higher welfare level in outcome *C*.

Confronted with a choice between only *A* and *B*, SCPA utilitarianism judges both equally permissible. The same holds for a choice between only *A* and *C*. However, if one is choosing between only *B* and *C*, the former is forbidden and the latter required. One could argue that this violates transitivity, since *B* is on a par with *A*, which is on a par with *C*, but *B* and *C* are not on a par. Such intransitivity cannot arise if one focuses only on the quantity of welfare that an outcome entails, as the impersonal view requires.

However, I argued that morality should be concerned with how individuals are affected and not with welfare as an abstract quantity. It should be concerned with net aggregate benefit. From this perspective, such intransitivity is not disturbing. As explained, whether an outcome entails a benefit or harm crucially depends on what the alternatives are. This follows from the standard, counterfactual comparative account of harm and benefit.[9]

Consider a case known as a *mere-addition paradox*, as depicted in figure 7.8. In outcome *A*, a population with very well-off individuals exists. In outcome

B, an additional population of somewhat less well-off individuals is added. In outcome *C*, the welfare level of both populations is equal: the original population is now slightly worse off than it was in outcome *B*, and the other population is now much better off than it was in outcome *B*.

The mere-addition paradox seems troubling. Intuitively, *B* is at least as good as *A*, and *C* is better than *B*. However, accepting that *C* is better than *A* would lead one to the repugnant conclusion. So, intuitively, one wants to resist this conclusion. But one cannot accept all these intuitive judgments without giving up on the (deontic) transitivity requirements (Ng 1989; Blackorby and Donaldson 1991; Arrhenius 2000).

SCPA utilitarianism rejects the (deontic) transitivity requirements and yields the intuitive judgments regarding this case. As Meacham points out:

> This also allows us to see the arguments offered by proponents of intransitivity, such as Temkin (1987), Rachels (1998) and Persson (2004), in a new light. Proponents of intransitivity can be seen as arguing that there is no "all things considered better than" relation which is (i) directly tied to moral obligation, (ii) situation-independent, and (iii) transitive. Proponents of person-affecting views [. . .] will agree. But proponents of intransitivity take the culprit to be (iii). Proponents of person-affecting views will take the culprit to be either (i) or (ii). I.e., either the "all things considered better than" relation is not directly tied to moral obligation (in which case it's of little interest), or it's not situation-independent. (2012, 21)

This shows that SCPA's rejection of the deontic transitivity requirement is not a weakness but a strength.[10]

Parfit's famous case of the miserable child, as depicted in figure 7.9, and similar cases present another challenge to saturating-counterpart person-affecting utilitarianism. Consider a couple that knows before conception that, perhaps due to some genetic defect, any child it could have would lead a miserable life, a life with negative welfare throughout (Parfit

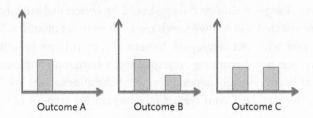

Outcome A   Outcome B   Outcome C

Figure 7.8 The Mere Addition Paradox

1984, 391). The couple has the choice between having and not having a child. Assume that there are no other relevant options.

According to both SCPA utilitarianism and prior-existence utilitarianism, there are no direct act-related reasons against conceiving the child. After all, the child is a contingent being, and existing as opposed to never existing cannot harm it. Impersonal utilitarianism, in contrast, clearly favors outcome A, because it contains more welfare.

While SCPA utilitarianism's judgment in this case may seem counterintuitive at first sight, on both views the child's welfare fully counts once it exists as a sentient being. If no better option is available, an act of abortion or euthanasia should then prevent the child's suffering. Knowing that this will be the case may, of course, provide an indirect reason against conceiving the child in the first place, because it may reduce the welfare of others. Furthermore, if a couple was planning to bring such a wretched child into existence and let it suffer, they could be condemned for having a bad character on utilitarian grounds (i.e. a character that tends not to maximize welfare). This move is possible if character is accepted as a direct evaluative focal point, along with actions (Louise 2006; Ord 2008; Sapontzis 1987). Remember that SCPA utilitarianism does not evaluate states of affairs as such. Instead, it tells us what is right or wrong. Conceiving the child as such is not a directly wrong action, unless it has negative effects on others. However, letting the child suffer once it exists is wrong, all else being equal. SCPA utilitarianism does not sanction letting the child suffer in any way.

Unfortunately, one can imagine situations in which it is impossible to end the individual's suffering or to end an individual's miserable life, once it exists. If there is no requirement not to bring such an individual into existence, the theory cannot condemn their suffering. If this were done intentionally, however, the theory could—as explained above—condemn the character of those who brought about such uncompensated suffering.

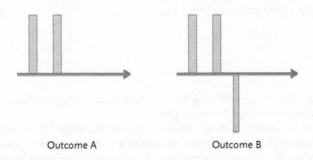

Outcome A   Outcome B

Figure 7.9 The Miserable Child

7.7. CONCLUSION

The presented theory accords with the idea that killing an animal and thereby depriving it of future welfare harms the animal. Bringing a second animal into existence that would not otherwise have existed does not benefit this second animal, because existence as opposed to non-existence cannot be a benefit (or harm). Since morality should be concerned with effects on sentient beings (i.e. with harms and benefits), the most plausible version of utilitarianism is one that requires maximizing not the quantity of welfare as such (as impersonal utilitarianism has it), but rather net aggregate benefit.

Different people cases—and different number cases in particular—have been widely discussed as the major challenge to person-affecting theories. A combination of person-affecting utilitarianism and a particular counterpart theory, as presented here, tackles these cases at least as successfully as rival theories do.

The resulting theory, which I call saturating-counterpart person-affecting utilitarianism, does not imply the replaceability argument. The version of utilitarianism presented here does not sanction routinely killing animals, not even if the animals lead pleasant lives and will be replaced by others whose lives contain as much welfare as the future lives of the killed animals would have contained and that would not otherwise have existed. I therefore conclude that utilitarians need not accept the replaceability argument. They can grant both human and nonhuman animals a stronger protection against killing.

ACKNOWLEDGMENT

I thank Christoph Fehige, Nils Holtug, and Robert Ranisch for helpful comments on earlier versions of this paper.

NOTES

1. Singer's theory allows keeping human babies for spare organs if this practice benefits people more than it enrages them.
2. As will be explained, Singer tried to restrict the scope of the replaceability argument to human and nonhuman animals that lack certain higher cognitive capacities. However, these restrictions do not seem to work, and Singer's current position does not allow for such restrictions.
3. See Luper (2007, 2009a) and Višak (2013, 85–90) for similar proposals.

4. The person-affecting view might better be called the "sentient beings-affecting view," because it is not restricted to harm and benefits for persons, but for sentient beings in general. Nevertheless, I will stick to the common label.
5. Meacham (2012, 267) adds a fourth condition, which will not interest us here, since it is irrelevant when applying his counterpart proposal to person-affecting utilitarianism.
6. Norcross (1999) brought forward these examples.
7. Cases in which the argument could be used to justify the "rearing and killing" of humans are conceivable as well.
8. For a list of defenders, see Matheny and Chan 2005.
9. Similar observations apply to the principle called "independence of irrelevant alternatives," which says that adding a third option should not change the relative ranking of two other options, and which can also be formulated as a deontic principle (Meacham 2012, 269).
10. Again, the same holds for SCPA utilitarianism's rejection of the deontic version of the principle of the independence of irrelevant alternatives (Meacham 2012, 269).

8

Singer on Killing Animals

SHELLY KAGAN

There are, I think, at least two questions that any adequate account of the ethics of killing animals should try to answer. First of all, and most importantly, we'd like to know whether it is indeed wrong (other things being equal) to kill animals at all. Of course, killing may often involve pain, and most of us would agree that it is wrong (again, other things being equal) to cause an animal pain. But recognizing this fact doesn't yet tell us whether there is anything wrong with killing the animal per se—that is to say, above and beyond the pain it might involve. Suppose that we are considering killing a given animal painlessly. Would that still be objectionable? If so, why?

Second, assuming for the moment that it is, in fact, wrong to kill animals, is there something particularly wrong about killing *people*? That is to say, if we distinguish between being a *person* (being rational and self-conscious, aware of oneself as existing across time) and being what we might call a "mere" animal (sentient, but not a person), we might wonder whether it is somehow worse to kill a person than it is to kill an animal that is merely sentient. Most of us, I imagine, think that something like this is indeed the case.[1] But it is not obvious whether this common view is justified, and even if it is, it is not obvious what makes the killing of the person worse. (Of course, many people think it clearly wrong to kill people and not at all wrong to kill a mere animal; if that's right, then it follows trivially that killing a person is worse. But it is far from clear that this common view is correct.)

There is a third question—less central than these first two, but philosophically fascinating nonetheless—that merits examination as well. If and when there would normally be something wrong about killing a given animal (whether a person or not), can that wrongness somehow be outweighed or canceled out by our creating a *new* animal, one that would not otherwise exist? If we *replace* the animal that is being killed, might that eliminate the objection to the killing that would otherwise be in order? Here, I suspect, most of us would want to put a tremendous amount of weight indeed on the distinction between persons and other animals. Many of us are at least tempted by the thought that merely sentient beings may be morally replaceable in just the way I have described. But it seems to be a different matter when it comes to people: people are *irreplaceable*. Or so most of us would want to insist. Once again, however, it is not at all clear whether this common view can be defended.

One of the most significant and influential discussions of all these matters can be found in Peter Singer's *Practical Ethics*.[2] Singer argues that it is indeed wrong to kill mere animals, though it is worse to kill people; and he defends the common view that people are irreplaceable, while mere animals are not. But the underlying philosophical issues are complicated, and in my own case, at least, I find that it isn't always easy to follow Singer as he tries to chart a course through the relevant philosophical thickets. Accordingly, in this paper I want to reconstruct the main lines of Singer's discussion, spelling out his view, and subjecting it to what I hope is friendly criticism. (To be somewhat more precise, I should say that what I want to examine is Singer's position as laid out in the third edition of *Practical Ethics*—for Singer readily admits to having changed his mind over the years, particularly with regard to our last question; and he remains uncertain about various aspects of that view.[3])

KILLING MERE ANIMALS

Singer begins his discussion of killing by examining at length our second question—the question of whether there is anything especially wrong with killing a person. But I propose to start instead with a version of our first question, asking whether there is anything wrong with killing a sentient creature per se. Suppose that we are dealing with a mere animal, sentient but not a person. And suppose as well that we are contemplating killing it painlessly (a qualification I won't keep repeating). What, if anything, would be wrong with doing that?

A natural and plausible suggestion is this: killing an animal that would otherwise have an overall pleasant future prevents it from experiencing the pleasure it would otherwise get. That's why killing it is wrong.

Oddly, though, while Singer thinks something like this must be right (pp. 85–86), he also thinks that saying this requires us to adopt a view according to which pleasure has objective value. We can't give the plausible answer, he believes, if we restrict ourselves to the kind of preference view to which Singer is strongly inclined, a view according to which what matters is simply furthering the preferences of those affected (pp. 12–13). After all, a merely sentient being has no *present* concern for its *future* pleasure. So if we kill it, there is no preference that gets frustrated. Accordingly, Singer thinks, if we are going to say that it is wrong to eliminate the pleasure that would have come to the animal, we have to ascribe objective value to that pleasure.

But I don't see why the preference theorist cannot avail himself of the obvious suggestion despite all of this. While it may be true that a merely sentient creature has no current preference for the future pleasure, it seems plausible to suggest, nonetheless, that if the pleasure *had* occurred, the animal *would* have had a relevant preference—namely, for the pleasant experience to continue (as Singer notes, p. 86). So there would have been a preference, and that preference would have been satisfied, had we not killed the animal. If the satisfaction of preferences is good, why doesn't that suffice to ground the wrongness of killing the animal in question?

Singer doesn't explain himself, but my best guess is that he is here unwittingly presupposing the truth of something like what he later calls "the debit model" of preferences (pp. 113–14), according to which there is no positive value in the satisfaction of preferences per se; there is only disvalue in their frustration. On such a view, the mere fact that the animal would have a satisfied preference in the future (a preference for the pleasure it would then be experiencing) does nothing to introduce any positive value; and so preventing this satisfied preference from ever arising does nothing to reduce the overall balance of value from what it otherwise would have been. Consequently, if we are going to insist that killing the animal eliminates something of value, we have to embrace objective value after all.

If this *is* what Singer has in mind (and it is only a guess), then I want to make three quick remarks. First, Singer is apparently presupposing a view here—the debit model—which he hasn't even mentioned yet (it is only introduced almost thirty pages later). Second, and more importantly, the debit model itself has seriously unintuitive implications (which Singer himself is at pains to bring out, pp. 113–17), so preference theorists might be well advised to reject it in any event. And finally, Singer is mistaken, I believe, in thinking that he himself needs to embrace the debit model (a point I'll explain later); so even Singer might do well to avoid it. In which case, it seems to me, the preference theorist is indeed entitled to offer the natural and plausible suggestion with which we started: killing a sentient creature is wrong (provided it will have an overall pleasant future) precisely because doing so prevents its future pleasure. As far as I can see, one simply needn't be an objectivist about value to say this.

As it happens, Singer also has a second worry about the natural suggestion (p. 87). If killing is wrong because of the loss of the potential future pleasure, by parity of reasoning don't we also have to say that it would be good to *create* extra sentient beings with pleasant lives? For that, too, would affect the total amount of pleasure (this time, in a positive direction). Yet many people find it implausible to hold that there is a significant moral reason to create extra sentient creatures merely because their lives would be pleasant ones.

This thought leads Singer to introduce some rival views concerning the ethics of creating new beings (pp. 88–90). He distinguishes between the "total view" (which accepts the implication in question, but which many find implausible for precisely this reason) and what he calls the "prior existence view" (which manages to avoid the implication, but which faces its own problems). But as far as I can see, Singer is mistaken in thinking that he needs to get into these issues at this point at all. For he is mistaken in thinking that the relevant counterpart to killing someone with a pleasant future is *creating* someone with such a future. Rather, it is *saving the life* of such a person.

After all, when we contemplate killing a sentient being, we are dealing with a being that *already* exists, not one that we will be bringing into existence. Accordingly, when asking what consistency commits us to here (with regard to aiding those with a pleasant future), we need only look at cases where the being we might aid already exists as well. Thus, if we insist—as we should—that it is bad to shorten a pleasant life (killing), all that consistency commits us to is the claim that it would be good to *lengthen* a pleasant life (saving). That is, if killing someone with a pleasant future is bad, then *saving* the life of someone with a pleasant future should be good. But this last judgment, of course, is one that virtually everyone will find intuitively acceptable.

As far as I can see, then, at this point in the discussion there is simply no need to get into the ethics of creating new beings at all. We need not ask whether it would be good to bring into existence a being with a pleasant future; we need only ask whether it would be good to *save* the life of a being who will then go on to have a pleasant future. And the answer, surely, is that this would indeed be good.

So the natural proposal stands. We can say that killing a merely sentient creature (with a pleasant future) is wrong because it robs the animal of the pleasant future it would otherwise have. The preference theorist can say this just as well as the objectivist about value. Furthermore, offering the natural proposal doesn't yet require a foray into the ethics of creation.

But this is not to say, of course, that we can avoid the ethics of creation altogether. Singer is right to bring it in, although I think he brings it in at the wrong place. Where we need it, rather, is when we turn to the topic of replaceability.

Suppose our choice is this: we can continue to raise a happy, merely sentient animal, or we can kill this animal but also replace it, by creating a new happy animal (of the same kind) that would not otherwise exist. (Imagine that we cannot simply add the second animal while still raising the first; perhaps we lack the resources to raise more than one at a time.) On the face of it, it seems, if killing the first animal would normally be wrong by virtue of the future pleasure that thereby gets eliminated, then this wrong will itself be avoided (or compensated for) by the creation of the second happy animal, since this reintroduces the same amount of pleasure as the killing eliminated, leaving no net loss. Apparently, then, killing a mere animal is not wrong in cases involving this sort of replacement. Merely sentient beings are replaceable.

I take it that something like this is Singer's understanding of the basic line of thought that might lead us to the conclusion that mere animals are replaceable. Of course, as Singer hastens to point out, even if this thought is correct it won't actually support anything like current factory farming practices (p. 106); but in principle at least—if there is no flaw in the argument—it does seem as though a mere animal should be replaceable. Accordingly, it is at just this point that we need to ask whether it is really true that creating a new animal with a pleasant life counts as introducing a good that can be used to compensate for the bad done by killing the first animal and preventing its future pleasure. If not, then the argument for replaceability can be blocked. Here, then, the ethics of creation is directly relevant.

But there is a different objection to the replacement argument that should be considered as well, and as far as I can see Singer doesn't notice it. Even if the pleasant life of the new animal does count as a good, and as great a good as the potential future pleasure that is destroyed by killing, it doesn't yet follow that it is morally permissible to kill the first animal. To think that it does follow is to assume that moral permissibility is a simple matter of adding up the good and the bad consequences, so that an act is morally permissible provided that none of the alternatives would have better consequences overall. It is precisely because Singer does assume this—because he thinks that mere animals are appropriately subject to a utilitarian calculus of this sort—that the permissibility of killing with replacement seems to be in play.

Note, however, that one might accept the initial thought that killing an animal is normally wrong because this prevents its future pleasure, and yet nonetheless reject the utilitarian treatment of animals, insisting, rather, that even mere animals should be handled within a *deontological* framework. Suppose, for example, that one were to hold that harm done to animals by robbing them of a pleasant future could *not* be justified by the mere fact that one will also bring about some further good. Then replaceability would not even be an issue: from a deontological point of view, the mere fact that an act's results may be good overall (or at least, not bad) simply won't suffice to justify doing harm.

Thus, those prepared to accord deontological rights to mere animals will not be moved by the possibility of replacement. And it does seem to me to be a shortcoming of Singer's discussion that he doesn't mention this sort of deontological approach here. Nonetheless, having noted this point I am going to put it aside, and I will follow Singer in thinking about these issues primarily from a utilitarian perspective.

Accordingly, we still need to ask whether it is really true that adding extra pleasant lives adds value. This is where the distinction between the total view and the prior existence view comes in. Singer notes, correctly, that both views have their counterintuitive implications (pp. 88–90). Indeed, I would want to add that pretty much every view here has some counterintuitive implication or the other.

Some jargon will be helpful. We're wondering what to say about cases where we face the possibility of creating an animal that would not otherwise exist. Call those animals who will exist *regardless* of the particular choice under examination the *inevitables* (since they will end up existing regardless of what choice I make now), and those animals whose existence depends on the particular choice being examined the *contingents*. Call those animals who actually end up existing at some time or the other—past, present, or future—the *actuals* (so all inevitables are actuals, but so are some contingents), and those animals that could have existed, but never do, *nonactual contingents*. If, for simplicity, we restrict our attention to the interests animals have in feeling pleasure and in avoiding pain, let us say that we *count* the pleasure a given animal would feel under some outcome if that gives us a reason to promote that outcome, while we count the pain they would feel if that gives us a reason to avoid that outcome. So our question becomes: Which interests are we to count? *Narrow* views count only the interests of the inevitables, while *wide* views count the interests of contingents as well; and *intermediate* views count the interests of *some* contingents, but not all of them (for example, an intermediate view might count the interests of all actuals, contingent or inevitable, but not the interests of nonactual contingents).

Armed with these distinctions, here are some possible positions (there are more):

- The *total* view is a wide view. It counts the interests of everyone who will or might exist. And it does this symmetrically, taking into account both pleasure and pain.
- The *prior existence* view is narrow, only counting the interests of the inevitables. But with regard to those interests, it symmetrically counts both pleasure and pain.
- *Actualism* is an intermediate view, counting the interests of all and only actuals. Thus it counts the interests of all inevitables (since they must be actual) and of all actual contingents, but it does not count the interests of nonactual contingents. It too treats pleasure and pain symmetrically, when they count at all.

In addition to these various symmetrical views, there are also *asymmetrical* views, the most important of which are wide. These count the interests of both inevitables and contingents, but they do so in a way that treats pleasure and pain asymmetrically. Here are some versions of asymmetry (there are others):

- *Extreme asymmetry*, according to which pain counts against an outcome, but pleasure does not count in its favor—not even the pleasure of inevitables.
- *Moderate asymmetry*, according to which both pleasure and pain count for inevitables, but not for contingents. For contingents, only pain counts, not pleasure.
- *Impure asymmetry* (asymmetry with an "actualist twist"), according to which the interests of all *actuals* are treated symmetrically (that is, both pleasure and pain count for actual contingents, and not only for inevitables); but for nonactual contingents only pain counts.
- *Offsetting asymmetry*, according to which pleasure can count to *offset* pain (sufficient pleasure can cancel out the reason to avoid a given outcome that would otherwise be generated by the existence of pain under that outcome) but pleasure can never count robustly in *favor* of an outcome (there would, for example, be no reason to promote an outcome with pleasure but no pain).

- *Impure offsetting asymmetry* (offsetting asymmetry with an actualist twist), according to which pleasure *can* count robustly in favor of an outcome, but only in the case of actuals. For nonactual contingents, pleasure can only be counted to offset pain; it provides no reason in its own right to favor an outcome.

These are, I think, the eight most prominent views concerning the ethics of creation. Unfortunately, as I have already suggested, none of these views are easy to accept.

Initially, to be sure, actualism seems rather attractive: what could be more reasonable than to count the interests of exactly those beings who will exist at some time or the other? (Surely, it seems, we shouldn't count the interests of nonactual contingents—merely potential beings that never actually exist!) But this view leads to implausible paradoxes. Suppose for example that we must choose between outcome 1, where A will be the only sentient being that exists, at a mildly positive level of well-being (+10), and outcome 2, where A will be somewhat better off (+20), but B will exist as well, with a life containing far more pain than pleasure (−100). If I create outcome 2, then both A and B are actual, so the interests of both count, in which case it would have been better to create outcome 1 instead (since A would be only somewhat worse off, while B would avoid having the miserable life). Yet if I create outcome 1, then only A is actual, so B's interests are irrelevant, in which case it would have been better to create outcome 2 instead (since A would be somewhat better off). So neither choice is acceptable! In effect, actualism ends up ranking outcomes in a way that implausibly depends on which choices we actually make.

Should we then accept the prior existence view? But this too gives an unacceptable answer in the case just discussed. When we face the choice between outcomes 1 and 2, A is the only inevitable, and so according to the prior existence view, A is the only being whose interests count. That means it is better to create outcome 2 (where A will be somewhat better off), even though this involves bringing into existence a sentient being—B—who will have a life of utter misery. According to the prior existence view, since B is a mere contingent her interests are irrelevant; but that seems unacceptable as well. (Singer makes a similar point with a similar case at p. 89; cf. pp. 108–11.)

The total view avoids this unacceptable implication, since it counts B's interest in avoiding the life of utter misery. It insists, plausibly, that we must choose outcome 1 rather than 2. Even though it is true that, if we avoid creating her, B remains a nonactual contingent, her interests still count. However, the total view also implies that we have reason to create extra sentient beings that will have pleasant lives. And as we have already noted, many find this implication implausible as well.

Asymmetry views can accommodate the intuition that there is no reason to create beings with pleasant lives, while still agreeing that there is reason to avoid creating beings with miserable lives. They do this, of course, by treating pleasure and pain asymmetrically. Roughly speaking, in the relevant cases they count the pain, but not the pleasure. But for exactly this reason such views strike most of us as unacceptably ad hoc (as Singer notes, p. 89). Absent an explanation for why pleasure should sometimes fail to generate a reason in cases where pain nonetheless would, asymmetry views seem philosophically unsatisfactory.

This vice—of being ad hoc—is common to all asymmetry views. But there are further specific objections to each particular asymmetry view as well. Extreme asymmetry has the unacceptable implication that there is no reason to make already existent sentient beings happier by increasing their pleasure. (Pleasure simply doesn't count, not even for inevitables.) Moderate asymmetry avoids this implication (since pleasures for inevitables do count), but it has the implausible implication that I act wrongly if I create a sentient being that has, overall, an incredibly pleasant life, as long as that being suffers any momentary pain whatsoever. (Recall that for moderate asymmetry, only the *pains* of contingents count, not their pleasures.)

Impure asymmetry avoids this last difficulty (since it counts the *pleasures* of contingent actuals, and not just their pains), but evaluations are once again implausibly dependent on one's choices (as with pure actualism). Admittedly, if I *do* create the being with the overall pleasant life, that is permissible (since the pleasure of the contingent actual counts and is more than sufficient to outweigh its pain). Yet if I do *not* create the being with the overall pleasant life, then—unacceptably—it turns out that it would have been *wrong* to create it! (For if I don't create this creature, it is a nonactual contingent; and for such beings pleasures don't count, so there is nothing to offset the pain, which does count.) Worse still, this view is doubly ad hoc: not only is it true that the *pain* of nonactual contingents counts, while the pleasure of nonactual contingents does *not* count, this very asymmetry only holds with regard to nonactual contingents. For actuals (whether contingents or inevitables), we have symmetry with regard to pleasure and pain instead.

From this perspective, offsetting asymmetry seems preferable. It avoids choice relative evaluation, and it says that it is permissible to create an animal with an overall pleasant life (since the pleasure offsets the pain). What's more, it isn't *doubly* ad hoc, since it treats pleasure as a mere offsetter for pain in *all* cases (not just for actuals). Nonetheless, this view does remain implausibly ad hoc (since we have no explanation of why pleasure should count only insofar as it offsets pain, while pain is not similarly restricted). And it implausibly fails to recognize that we can have good reason to *increase* the pleasure of an already existing being—even if the animal already has an overall pleasant life. (On this view, one only has reason to increase the pleasure of existing animals up to the point where they reach the neutral level and all of the pain has been countered.)

Finally, while impure offsetting asymmetry avoids this last problem and recognizes that we have reason to increase the happiness of animals with lives that are already pleasant overall (since on this view pleasure counts robustly—and not merely as an offsetter for pain—in the case of actuals), it does this at the cost once more of being doubly ad hoc (like impure asymmetry): not only is it true, when it comes to nonactual contingents, that pleasure remains a mere offsetter (while pain is not similarly restricted), this asymmetry holds *only* with regard to nonactual contingents, and not at all for actuals. And of course—a last objection—the actualist twist behind impure offsetting asymmetry reintroduces the unacceptable dependence (already noted for actualism and impure asymmetry) of evaluations on one's actual choice.

As we can see, then, all the main views on this vexed topic are problematic. I've belabored this point because it should make it easier to recognize a potentially surprising fact, that the *correct* view—whatever it is—will inevitably have one or more features that will strike at least some of us as rather implausible.

So what *is* the correct view? Reasonable people, no doubt, can disagree about this; but for my money, at least, I am inclined to accept the *total* view. On balance, it seems to me, this is the least unacceptable of the various alternatives.[4]

If we do accept the total view—and if we keep in mind our earlier decision to follow Singer in thinking about mere animals within a utilitarian framework—then replaceability for mere animals does seem to follow. If I kill an animal with a pleasant life, I prevent the pleasure it otherwise would have had. But if I replace it with a new animal with an equally pleasant life, an animal that would not otherwise exist at all, then according to the total view this newly created pleasure counts as well, and by hypothesis it is as great a good as the pleasure that is prevented. So the result is no worse, overall, than had I simply allowed the original animal to live (and never created the replacement). Given a utilitarian framework, this means that my complex act—of killing and replacing—is morally permissible.

In short, given our assumptions, while killing a mere animal with an overall pleasant future is normally wrong, it will not be wrong if we replace that animal with another one. Merely sentient beings are morally replaceable.

KILLING PEOPLE

In the previous section I defended the claim that it is wrong to kill mere animals (with pleasant lives) on the ground that doing so robs them of the pleasure they would otherwise have. If sound, presumably this same line of thought applies to *people* as well. After all, nothing in that earlier argument turned on the idea that the animals in question were *merely* sentient beings, as opposed to being both sentient *and* rational and self-aware. So at a minimum, if some person would otherwise have a pleasant life, to kill her would normally be wrong, at least in part precisely because doing so would rob her of the pleasure she would otherwise have. Unsurprisingly, then, it is wrong to kill people, and not only mere animals.

But might there be something especially wrong with killing a person? Might this be wrong in some way that goes beyond the wrongness involved in killing any animal at all with a pleasant future? Singer thinks so, and mentions four possible considerations (pp. 76–85). All turn on the fact that since people are rational and self-conscious, aware of themselves as existing across time, someone with a pleasant future will typically have a desire that he continue to exist *into* the future. (Accordingly, these four considerations won't apply to the killing of mere animals.)

First of all, then, if I kill some person, this may well affect others who themselves have preferences about their continued existence (pp. 76–80). Killing one person creates anxiety (a kind of pain) in others, and this contributes to the wrongness of that killing. Now this first consideration certainly does seem relevant to the morality of killing; but since it is a highly indirect reason (and it depends as well on the possibility of being found out), Singer doesn't emphasize it.

The second consideration focuses on the preference of the victim himself (pp. 80–81). If—as is typically the case—the person wants to continue to exist, killing this person frustrates his preference. Indeed, it will normally frustrate many of the person's deepest and most central preferences. Therefore, if we accept the preference view—according to which it is wrong, other things being equal, to frustrate the satisfaction of a preference (and the stronger the preference, the greater the wrong)—then it will normally be a great wrong to kill a person. It is this consideration, I think it fair to say, that Singer finds the most compelling.

But Singer mentions two other considerations as well. Third, then, it might be that people have a *right* to life, grounded in their desire to continue to exist into the future (pp. 81–83). Singer emphasizes the thought that having such a desire may be a necessary condition for having the right in question; so mere animals will lack this right. (Perhaps this explains why—as I noted earlier—Singer does not explore the possibility of treating mere animals within a deontological framework.) Finally, Singer notes that respect for autonomy may generate *further* reason not to kill a person, since killing a person typically interferes with their autonomous choice to go on living (pp. 83–84).

Although there are various subtle differences between the second consideration and the last two, all three can be seen as giving expression to the thought that if a person wants to continue living (if this is her preference, desire, or choice), then other things being equal this very fact makes it wrong to kill her. The most important place where they differ, it seems to me, is with regard to the *strength* of the reason in question not to kill. Singer understands the preference view in terms of utilitarianism, and as such the second consideration can, in principle, be outweighed, if conflicting preferences of others come into play. In contrast, the third and fourth views take the reason not to kill to be stronger—"more absolute," as Singer puts it—less readily outweighed by "utilitarian calculation" (p. 81). In short, the third and fourth approaches express *deontological* perspectives (though Singer doesn't use the term). They are intended to capture the view, held by many, that killing a person cannot be justified even when greater good might come of it.

This point is important, because it seems to me that Singer loses sight of it. Although he tells us that we should keep all four views "in mind" (p. 85), by the time Singer turns to a discussion of whether people are replaceable (as mere animals seem to be), the third and fourth views are forgotten; Singer considers the issue only from the perspective of the preference utilitarian. That discussion is certainly not without its interest. But it is a shortcoming nonetheless that Singer fails to so much as consider the issue from a deontological perspective as well.

Having noted this point, however, I am once again going to put it aside. Let us follow Singer and ask ourselves what a *preference utilitarian* should say about the replaceability of people. Normally, as we have seen, it is wrong to kill a person. But might it be permissible to kill someone *provided* that one also creates a *new* person (with an equally pleasant life) who would not otherwise exist?

Most of us, presumably, would insist that the answer is no. But Singer worries that the preference utilitarian may not be in a position to say this (p. 113). After all, while it is true that if we kill the first person his preference to remain alive is frustrated, the new person will *also* have a preference to be alive—and *that* preference *will* be satisfied. Since we end up with a satisfied preference either way, regardless of whether we kill and replace or simply let the first person be, the result seems to be no worse if we do kill and replace. So the preference utilitarian apparently has no ground for objecting. It seems, then, that according to preference utilitarianism not only are mere animals replaceable, people are replaceable too!

Singer takes this objection very seriously, and he significantly complicates his view so as to avoid the unwanted implication. One thing he does—as I have already noted—is to embrace the debit model of preference satisfaction (pp. 113–14). According to this view, recall, there is no positive value in the satisfaction of a preference, there is only negative value in the existence of unsatisfied or frustrated preferences. That is, although satisfying a preference eliminates or avoids the bad constituted by frustrated preferences (and so is good in this minimal, comparative sense), it introduces no robust, positive good in its own right. (Incidentally, although Singer explicitly endorses the debit model only with regard to *creating* preferences, as far as I can tell he means to accept it across the board, even with regard to already existing preferences.)

If the debit model is correct, then the argument for replaceability is blocked. When we kill the first person, this frustrates his desire to go on living, which is bad; and the fact that we also create a new person with a new preference—a preference that is satisfied—is simply irrelevant, since the satisfaction of that preference isn't good, but merely avoids a potential further bad (the bad that would occur if that new preference were unsatisfied). Thus, if I kill and replace, the result is actually worse (since the victim has a frustrated preference) than if I had merely let the first person be (for then his preference would be satisfied). In short, given the debit model, people are not replaceable.

Should we accept the debit model? Singer notes that it accommodates a certain number of our intuitions (pp. 113–14), but, as he also notes, this view also has an implication that most of us will find utterly unacceptable (p. 114): given the debit model and preference utilitarianism, it is wrong to have any children *at all*—no matter how happy they may be—since everyone has at least some unsatisfied preferences, however mild. After all, even the child's vast array of satisfied preferences can do no more than cancel out *most* of the "debit" thereby created, not all of it. So having a child must be wrong. (Similarly, on this view, a world with no sentient creatures whatsoever would be better than an otherwise wonderful world in which anyone at all has *any* unsatisfied preferences; cf. p. 116.)

To avoid this latest implication, Singer complicates his view yet again (pp. 116–18). He entertains the possibility that pleasure is objectively good—not dependent for its value on being the object of preference. Indeed, he entertains the possibility that there may be other objective goods as well. If there *are* any such objective goods, then the objection can be answered: when we bring the overwhelmingly happy child into existence, the objective value of the pleasure, say, that she has (along with other objective goods, perhaps) outweighs the bad of having some unsatisfied preferences. So having the happy child remains permissible. (A similar answer, of course, applies to the example of the wonderful world.)

Unfortunately, Singer doesn't seem to notice that introducing objective goods into his theory reopens the door for the argument for replaceability. After all, if I kill and replace, it may be true that the satisfied preference of the new person doesn't count as a robust positive (given the debit model), but for all that, there will still be *objective* goods that are created as well (for example, the pleasure of the new person). Mightn't these outweigh the bad generated by frustrating the preference of the person we kill?

Since Singer doesn't discuss this worry, it isn't clear how he would reply to it. But perhaps we can suggest the following on Singer's behalf: Admittedly, when we kill and replace this will introduce some objective value, which may outweigh the bad due to the frustration of our victim's preference to live. But since, by hypothesis, the first person would have had a life as valuable as his replacement, it too would have contained objective goods (had he not been killed) of the same sort and significance. So the results are indeed better if we *avoid* killing: we have the same amount of objective value, while we avoid frustrating the preference of our potential victim.

Arguably, then, if he embraces both the debit model and the existence of objective value, Singer can indeed avoid the conclusion that people are replaceable. Nonetheless, it seems to me that there was never any need for Singer to introduce these various complications in the first place! For Singer erred, I believe, in thinking that preference utilitarianism straightforwardly implies replaceability. On the contrary, as far as I can see, given future-regarding preferences of the sort we are already entertaining, replacement will not normally involve as good a result as refraining from killing. Put simply, Singer failed to take into account some of the relevant preferences.

Here is a simple example to demonstrate the idea. First, let's adopt a fairly straightforward (if overly simple) version of preference utilitarianism: For each frustrated preference we assign a score of -1; and for each satisfied preference we assign a score of +1. (Recall, we are looking to see if the preference theorist is forced to embrace the debit model, so we are here doing without it. Accordingly, satisfied preferences are assumed to have positive value.) Next, suppose (again, for simplicity) that the normal life of a person consists of four moments or stages. Imagine that at each such moment a person will normally have a preference to be experiencing pleasure at that moment. But more than this, at any given moment, and for each future moment (of the four), a person will normally have a preference to experience pleasure at that later moment as well. Then at the first moment, t1, the person will have four relevant preferences: he will have a preference to be experiencing pleasure at t1, a preference to be experiencing pleasure at t2, a preference to be experiencing pleasure at t3, and a preference to be experiencing pleasure at t4. In contrast, at the second moment, t2, the person will have only three relevant preferences (one each for t2, t3, and t4). At the third moment, t3, the person will have two relevant preferences (one each for t3 and t4); and at the final moment, t4, the person will have one relevant preference (a preference with regard to t4). Finally, suppose that, if left alone, the person will experience pleasure at each of his four moments.

Given all of this, it is easy to see that if he is left in peace, all of the person's preferences will be satisfied. And since there are ten of them in total (4 in the first moment, 3 in the second, 2 in the third, and 1 in the last), this results in an overall score, with regard to this life, of +10.

Now let us compare this score to the score we get over the *same four moments* if the first person is killed after t2, and then immediately replaced with someone new, who will have her life begin at t3. In this case, the first person has only three of his preferences satisfied (his preferences at t1 with regard to t1 and t2, and his preference at t2 with regard to t2), while four of his preferences are frustrated (his preferences at t1 with regard to t3 and t4, and his preferences at t2 with regard to t3 and t4). So far, then, this gives us a score of 3 + -4 = -1. (Obviously enough, since he is killed after t2 he never comes to have any preferences at t3 or t4.)

And what about the replacement? Let us suppose that she is allowed to live out a normal life of four moments. Still, if we are only interested in the value that arises during t1–t4, then we should only look at the preferences of the replacement that exist during t3 and t4 (disregarding those that exist later, during t5 and t6). There are seven such preferences, all satisfied, for a score of +7. Adding this to the score of -1 generated by the satisfied and frustrated preferences of our first person, this gives a total score for the killing and replacement scenario of -1 + 7 = +6 for the period in question (t1–t4).

In contrast, as we have already noted, had we let the first person live out his natural lifespan, the total score during the same period would have been +10! In short, even according to the preference utilitarian, killing a person and replacing him leads to worse results overall, during the course of what would have been the first person's life.

This example is, of course, unrealistically simplified; but not, I think, in ways that affect the main point.[5] Given the fact that people ordinarily have preferences with regard to their future, killing someone with a pleasant life and replacing him will normally have worse results—over the same initial period of time—from the standpoint of preference utilitarianism. So it is not true in any straightforward sense that under preference utilitarianism it is just as good to kill someone and replace them as it is to let them be. Unlike mere animals, people are not replaceable in that way.

To be sure, it might be objected that if we take into account *all* of the results—not just those that occur during what would have been the first person's lifetime, but also those that occur afterward—preference utilitarianism might well sometimes conclude that the results would be better if we did kill and replace. In principle, after all, the satisfied preferences that the replacement person will have *later* in life might well outweigh the frustrated preferences of the victim. (That won't be true, as it happens, in our particular example; but it certainly can be true if the victim is killed late enough in life, and the replacement lives for a long enough time afterward.)

But that still doesn't mean that the preference utilitarian thinks it permissible to kill and replace. For if we are going to extend the time frame of our comparison, and take into account the benefits from replacement due to the increase in the total amount of life that replacement makes possible (that is to say, take into account the gains from satisfied preferences that occur *after* t4), then it is important to bear in mind that replacement will provide even *greater* overall gains if it is done without killing. That is, the results would be better still if, instead of killing first and then replacing, we create the new person only *after* the first person has lived out his life. Letting live and *then* replacing will normally have better results than *killing* and replacing. So even when we calculate over the extended time frame, simple killing and replacement won't normally be permissible.

In short, the preference utilitarian won't normally approve of killing a person, even *with* replacement. The results of killing will normally be significantly worse than the results we would have under alternative courses of action that don't involve killing at all.

Where then does that leave us? None of this should be taken to show that preference utilitarianism gives acceptable answers across the *entire* range of cases that involve killing (with or without replacement).[6] But it does show, I think, that it is a mistake to suggest that preference utilitarianism straightforwardly implies that people are replaceable. At a minimum, the question is far more complicated than one might initially have thought. But more than this, I am inclined to think that we are justified in accepting a considerably stronger conclusion: the preference utilitarian does not view people as replaceable at all.

If I am right, then perhaps Singer never needed to complicate his position in the various ways that he did. Or rather, somewhat more precisely, he didn't need to do it so as to avoid saying that people are *replaceable*. For all that we have seen so far, Singer could have remained a simple preference utilitarian after all.[7]

NOTES

1. However, many people would also want to grant human infants and the severely cognitively impaired a status similar to (or identical with) that of being a person, even though these humans are not in fact persons in the technical sense I have just characterized. Those sympathetic to this view sometimes appeal to the given human's *potential* to become a person, or to their *membership in a species* whose typical adult members are persons (or perhaps, in some cases, to their *having been* a person). Whether anything like this could suffice to earn a nonperson a moral status similar to that of someone who actually *is* a person, is a complicated and controversial question that, for simplicity, I want to put aside here. In what follows, therefore, assume that the "mere animals" I am discussing are ones that lack any such relevant connection (if such there be) to personhood.
2. Singer 2011. All parenthetical citations in this paper are to this work.
3. Indeed, Singer appears to have changed his mind yet again about some of the ideas discussed below. (See *The Point of View of the Universe* [Lazari-Radek and Singer 2014], published after this paper was completed.) Regardless, the discussion in *Practical Ethics* remains significant, whether or not Singer still accepts all of the terms in which it is framed.
4. For a contrary view, see Višak 2013.
5. For example, we might prefer to think of time as continuous (rather than discrete), with individual preferences lasting more than one moment. But a preference theorist can then hold that the value of satisfying a preference (or the disvalue of frustrating it) is proportional to the length of time that it is held. This will generate the same results as the simple model discussed.
6. The argument I have given assumes that one's future-regarding desires are *indexical*, rather than qualitative: I want it to be the case that *I* am happy—not merely that some creature be happy, nor even that some creature that thinks he is Shelly Kagan be happy. Presumably, of course, this is normally a realistic assumption. But imagine that in some case I want only that someone with my various memories, beliefs, and goals (and so on) exist in the future. Then certain types of killing and replacement—perhaps involving the creation of an exact qualitative duplicate of me—may turn out to be acceptable after all. That seems to me to be the right position to take, but I won't pursue the question here.

7. This is not to say that I think that preference utilitarianism *is* the best position to take. I myself would rather embrace a more objectivist view about value. For example, while Singer is right to suggest (pp. 90–93, 103–4, 122) that the lives of persons have greater value than the lives of mere animals—particularly those persons whose lives are richly interwoven due to a wide range of cross-temporal desires covering significant stretches of time—I would want to explicitly add what Singer does not, that this is a form of *objective* value. Although Singer's (hesitant) turn to objectivism may not be required for the reasons he thinks it is, I believe he is right to take it nonetheless.

9

A Kantian Case for Animal Rights

CHRISTINE M. KORSGAARD

9.1. INTRODUCTION

Kantian moral philosophy is usually considered inimical both to the moral claims and to the legal rights of nonhuman animals. Kant himself asserts baldly that animals are "mere means" and "instruments," and as such may be used for human purposes. In the argument leading up to the second formulation of the categorical imperative, the Formula of Humanity as an end in itself, Kant says:

> Beings the existence of which rests not on our will but on nature, if they are beings without reason, have only a relative worth, as means, and are therefore called *things*, whereas rational beings are called *persons* because their nature already marks them out as an end in itself, that is, as something that may not be used merely as a means. (G 4:428)[1]

In his paper "Conjectures on the Beginnings of Human History," a speculative account of the origin of reason in human beings, Kant explicitly links the moment when human beings first realized that we must treat one another as ends in ourselves with the moment when we realized that we do not have to treat the other animals that way:

> When [the human being] first said to the sheep, "the pelt which you wear was given to you by nature not for your own use, but for mine" and took it from the sheep to wear it himself, he became aware of a prerogative which, by his nature,

he enjoyed over all the animals; and he now no longer regarded them as fellow creatures, but as means and instruments to be used at will for the attainment of whatever ends he pleased. (CBHH 8:114)[2]

In his account of legal rights, Kant introduces a further difficulty for the cause of animal rights. For Kant, the point of legal rights is not, as many philosophers have supposed, to protect our more important interests. Rather, it is to define and uphold a maximal domain of individual freedom for each citizen, within which the citizen can act as seems just and good to him. In John Rawls's language, it is to create a domain in which each person can pursue his own "conception of the good."[3] Kant believed that each of us has an innate right to freedom, which he defined as "independence from being constrained by another's choice" (MM 6:237). He argued that without the institution of enforceable legal rights, our relationships with each other must be characterized by the unilateral domination of some individuals over others. The problem is not, or not merely, that the strong are *likely* to tyrannize over the weak. Even if the strong were scrupulous about not interfering with the actions or the possessions of the weak, without rights the weak would be able to act on their own judgment and retain their own possessions only on the sufferance of the strong (MM 6:312). Since her innate right to freedom is violated when one person is dependent on some other person's good will, Kant thinks it is a duty, and not just a convenience, for human beings to live in a political state in which every person's rights are enforced and upheld (MM 6:307–8).[4] No matter how well-intentioned we are, we can be rightly related to each other only if we live in a political state with a legal system that guarantees the rights of everyone.

But nonrational animals apparently do not have the kind of freedom that rights, on this account, are intended to protect. It is because human beings are rational beings that we are able to choose our own way of life. Rationality, for Kant, is not the same thing as intelligence. It is a normative capacity, grounded in what Kant took to be the unique human ability to reflect on the reasons for our beliefs and actions, and to then decide whether they are good reasons or bad ones. As rational beings, we reflect on what counts as a good life, decide the question for ourselves, and live accordingly. In the liberal tradition, with its strong emphasis on toleration and its antagonism to paternalism, this kind of autonomy has often been regarded as the basis of at least some of our rights. We have the basic rights of personal liberty, liberty of conscience, and the freedom of speech and association, because each of us has a general right to determine for ourselves what counts as a worthwhile life, and to live that life, so long as the way we act is consistent with upholding the same right for everyone else.

But Kant extends this account to all of our rights. He thinks that we must have property rights, for example, because if we did not, no one could use natural objects—a piece of land to grow crops on, for example—to pursue his own projects without being dependent on the willingness of others not to interfere with that use. Our right to property is therefore not grounded directly in our interests, but rather is seen as an extension of our freedom of action. Of course, Kant thought that one of the things in which we could claim property is the other animals. Their legal status as property is the direct correlate of their moral status as mere means.

Grounding all of our rights in freedom is important to Kant, because on Kant's account, rights, by their very nature, are coercively enforceable. It is the essence of having a right that you may legitimately use force to protect that to which you have the right, or the state may do so on your behalf. That is how rights secure our freedom against the domination of others. Kant believed that the protection of freedom is the *only* thing that justifies the use of coercion, because the protection of freedom is the use of coercion against coercion itself. According to Kant, people do not get to push each other around in the name of what one or another of us, or the majority of us, or for that matter even all of us, considers to be good. The only thing that justifies us in preventing someone from acting as she chooses is that her action is a hindrance to someone else's freedom.

But the other animals are not autonomous and do not choose their own way of life. This seems to imply that, in Kant's legal philosophy, questions about the rights of nonrational animals cannot even come up. And, of course, those who champion rights for animals are not usually interested in securing their freedom of action, but rather in securing them protection from harm. This seems to suggest that Kant's philosophy is not the place to look for a philosophical foundation for animal rights.

Nevertheless, in this paper I will argue that a case for both the moral claims and the legal rights of nonhuman animals can be made on the basis of Kant's own moral and political arguments. Kant's views about the human place in the world—his resistance to the pretensions that human beings have metaphysical knowledge of the way the world is in itself, and the arguments he uses to show that we can construct an objective moral system without such knowledge—require us to acknowledge our fellowship with the other animals.

9.2. WHY WE MUST REGARD ANIMALS AS ENDS IN THEMSELVES

In the argument leading to the Formula of Humanity, as I mentioned earlier, Kant claims that the nature of rational beings or "persons" "marks us out" as ends in ourselves. As some people read this argument, Kant is simply making a metaphysical claim about a certain form of value. Rationality or autonomy is a property that confers a kind of intrinsic value or dignity on the beings who have it, and therefore they are to be respected in certain ways. Lacking this property, the other animals lack this dignity or value.

There are several problems with understanding Kant's argument this way. One is that it does nothing to explain the particular *kind* of value that rational beings are supposed to have. "Value" is not a univocal notion—different things are valued in different ways. The kind of value that Kant thinks attaches to persons is one in response to which we respect their choices, both in the sense that we leave people free to determine their own actions, and in the sense that we regard their chosen ends as things that are good and so worthy of pursuit. This is made clear by the nature of the duties that Kant thinks follow from the injunction to respect persons as ends in themselves (G 4:429–31). We are obligated not to usurp other people's control over their own actions by forcing or tricking them into doing what *we* want or think would be best—that is, we are not allowed to use other people as mere means to our ends. We also have a duty to promote the ends of others. A person could certainly have some kinds of value—even some kinds of value as an end—without it following that his choices ought to be respected. A prince, or someone held by some religious tradition to be the embodiment of their god, might be valued the way a precious object is valued—preserved and protected and cherished—without ever being allowed to do anything that he chooses.

But the more important problem is that the proposed claim about the intrinsic value of rational beings is exactly the sort of metaphysical claim whose pretensions Kant's philosophy is designed to debunk. Kant does not believe that human beings have the kind of direct rational insight into the nature of things that might tell us that certain entities or objects are, as a matter of metaphysical fact, intrinsically valuable. Speaking a bit roughly, Kant thinks that claims that go beyond the realm of empirical or scientific knowledge must be established as necessary presuppositions of rational activity—that is, as presuppositions of thinking in general, or of constructing a theoretical understanding of the world, or of making rational choices. His philosophical strategy is to identify the presuppositions of rational activity and then to try to validate those presuppositions through what he calls "critique."[5]

In his argument for the Formula of Humanity, Kant aspires to show us that the value of people as ends in themselves is a presupposition of rational choice. The argument, as I understand it, goes like this:[6] Because we are rational, we cannot decide to pursue an end unless we take it to be good. This requirement is essentially built into the nature of the kind of self-consciousness that grounds rational choice. A rational being is one who is conscious of the grounds on which she is tempted to believe something or to do something—the purported reasons that move her to adopt a belief or an intention. Because we are conscious of the grounds of our beliefs and actions, we cannot either hold a belief or perform an action without endorsing its grounds as adequate to justify it.[7] To say that the pursuit of an end is justified is the same as to say that the end is good (C2 5:60). Importantly, Kant takes the judgment that the end is good to imply that there is reason for *any* rational being to promote it. As he says in the *Critique of Practical Reason*:

> What we are to call good must be an object of the faculty of desire in the judgment of every reasonable human being, and evil an object of aversion in the eyes of everyone. (C2 5:61)[8]

What he means is not that everyone must care about the same things that I do, but rather, that if my caring about an end gives me a genuine reason for trying to make sure that I achieve it, then everyone else has a reason, although of course not necessarily an overriding one, to try to make sure that I achieve it as well.

Consequently, Kant envisions the act of making a choice as the adoption of a certain "maxim" or principle as a universal law, a law that governs both my own conduct and that of others. My choosing something is making a law in the sense that it involves conferring a kind of objective—or, more properly speaking, intersubjective—value on some state of affairs, a value to which every rational being must then be responsive. It is important to Kant's own understanding of the implications of this argument that it is only rational choices that have this normative character. Only rational choices are made on the basis of an assessment of the grounds or reasons for them, and so only *rational* choices represent decisions about what *should* be done. The other animals do not make choices in the same sense that rational beings do, and such choices as they do make do not have the character of laws.

Most of the ends we choose, however, are simply the objects of our inclinations, and the objects of our inclinations are not, considered just as such, intrinsically valuable. As Kant puts it:

The ends that a rational being proposes at his discretion as effects of his actions (material ends) are all only relative; for only their mere relation to a specially constituted faculty of desire gives them their worth. (G 4:428)

The objects of your own inclinations are only—or rather at most—good *for* you; that is, good relative to the "special constitution" of your faculty of desire.[9] As Kant thinks of it, they are, usually, things that you like and that you think would make you happy. Now it does not generally follow from the fact that something is good *for* someone in particular that it is good absolutely, and that anyone has reason to promote it. As I have already mentioned, Kant supposes that a rational being pursues an end only if she thinks it is good absolutely, so he thinks we do not pursue the objects of our inclinations merely because we think those ends are good for us. Yet we *do* pursue the objects of our inclinations, and we often expect others to help us in small ways, or at least not to interfere without some important reason for doing so. That suggests that we take it to be absolutely good that we should act as we choose and get the things that are good for us. Why do we do that?

That is the question from which the argument for the Formula of Humanity takes off, and Kant's answer is that we do it because we take ourselves to be ends in ourselves:

Rational nature exists as an end in itself. The human being necessarily represents his own existence this way; so far it is thus a *subjective* principle of human actions. (G 4:429)

We "represent" ourselves as ends in ourselves insofar as we take what is good for us to be good absolutely. It is as if whenever you make a choice, you said, "I take the things that are important to me to be important, *period*, important absolutely, because I take *myself* to be important." So in pursuing what you think is good for you as if it were good absolutely, you show that you regard yourself as an end in itself, or, perhaps to put it in a better way, you *claim* that standing. Kant then continues:

But every other rational being also represents his existence in this way consequent on just the same rational ground that also holds for me; thus it is at the same time an *objective* principle. (G 4:429)

Kant tells us that at this point in the argument that is just a "postulate," which he will prove later in the book, in its final section. In the final section of the book, Kant sets out the grounds that he thinks validate our conception of ourselves, considered as rational beings, as members of what he calls a Kingdom of Ends, a community in which all rational beings as ends in themselves together make laws for themselves and for one another whenever they make choices.

So whenever you make a rational choice, then, you presuppose that you (and by implication, every other rational being) have a kind of normative standing, the standing of a legislator in the Kingdom of Ends, whose choices are laws to all rational beings. It is in this sense that Kant thinks your rational nature "marks you out" as an end in itself. Of course, in the moral realm, your right to confer objective value on *your* ends and actions is limited by everyone else's right to confer objective value on *his* ends and actions in the same way. (This is analogous to the way that, in the political realm, your freedom is limited by the like freedom of everyone else.) So only if your principle or maxim is morally permissible does it really count as a law. In Kant's own language, your maxim must conform to the categorical imperative: you must be able to will it as a universal law. Kant takes that to means that ultimately it is a rational being's capacity for *moral* choice that "marks him out" as an end in himself:

> Now morality is the condition under which alone a rational being can be an end in itself, since only through this is it possible to be a lawmaking member in the kingdom of ends. (G 4:435)

While recounting these arguments, I have switched back and forth between talking about our standing as lawmakers and talking about our standing as beings whose ends and actions should be regarded as good, and so as normative for everyone. That reflects the fact that there are two slightly different senses of "end in itself" at work in Kant's argument, which we might think of as an active and a passive sense. I must regard you as an end in itself in the active sense if I regard you as capable of legislating for me, and so as *placing* me under an obligation to respect your choices or to help you to pursue your ends. I must regard you as an end in itself in the passive sense if I am obligated to treat your ends, or at least the things that are *good for you*, as good absolutely. Kant evidently thinks that these two senses come to the same thing, for in his most explicit statement about why we have duties only to rational beings, Kant says:

> As far as reason alone can judge, a human being has duties only to human beings (himself and others), since his duty to any subject is moral constraint by that subject's will. (MM 6:442)[10]

But that does not obviously follow. The idea that rational choice involves a presupposition that *we* are ends in ourselves is not the same as the idea that rational choice involves a presupposition that *rational beings* are ends in themselves, for we are not merely rational beings. The *content* of the presupposition is not automatically given by the fact that it is rational beings who make it. Do we presuppose our value only insofar as we are beings who are capable of willing our principles as laws? Or do presuppose our value as beings *for whom* things can be good or bad? In fact, Kant's argument actually shows that we presuppose our value as beings for whom things can be good or bad—or, as we might put it for short, as beings who have interests. Let me explain why.

Suppose I choose to pursue some ordinary object of inclination, something that I want. According to Kant's argument, this choice presupposes an attitude I have towards myself, a value that I set on myself, or a standing that I claim. Is it my value as an autonomous being capable of making laws for myself as well as other people? Or is it my value as a being for whom things can be good or bad?

If it is the value that I set on myself as an autonomous being, then when I make a choice, I should be motivated by respect for my own autonomy, my capacity to make laws. The natural way to understand the idea that I respect my own autonomy is to suppose that I conform to a law simply because I myself have made it. Kant certainly thinks that whenever I make a choice I make a kind of law for myself, as well as for other people, and the idea is not without content; indeed, it is the essential difference between choosing something and merely wanting it. Wanting something, which is just a passive state, does not include a commitment to continuing to want it, but willing something, which is an active state, does include a commitment to continuing to will it, everything else being equal.

For example, if I choose (or will, in Kant's language) to grow vegetables in my garden, knowing that this will require me to weed it on a regular basis, then I commit myself to weeding my garden at certain intervals in the future, even if it should happen that I do not feel like doing so. This is not to say that I decide that I will weed my garden no matter what—though the heavens fall, as it were. But it is to say that when I take something as the object of my will or choice, it follows that any good reason I have for abandoning this object must come from other laws that I have made or other commitments that I have undertaken, and not merely from a change in my desires. Having willed to grow vegetables in my garden, I can decide not to weed it if I need to rush to the bedside of an ailing friend, for instance. But I have not really decided, or willed, to grow vegetables in my garden if I leave it open that I will not weed my garden if I just do not happen to feel like it. For if *all* that I have decided when I decide I will keep my garden weeded is that I will weed it if I happen to feel like it, then I have not actually decided anything at all.[11]

So when I choose to grow vegetables as my end, I bind my future self to a project of regular weeding by a law that is not conditional on my future self's desires. In that sense, I have legislated a categorical imperative for myself. But my future self, in turn, also binds me, for it is essential that if she is going to do the necessary weeding, I must now buy some pads to protect her knees, and the tools for her to weed with—and I must also do *that* whether I feel like it or not. In this simple sense, when I make a choice, I impose obligations on myself—I create reasons for myself. When I act on those reasons, you can say that I am respecting my own autonomy, by obeying the law that I myself have made.

When someone else respects my choice, he is also governed in this way by respect for my autonomy: he takes my choice to be law. But my own *original* decision to choose or will some desired end is not motivated by respect for my own autonomy in *that* sense. I cannot respect my own choice or do what is necessary to carry it out until *after* I have made that choice. So the sense in which I "represent myself" as an end in itself when I make the original choice is not captured by the idea that I respect my own autonomy, in the sense of taking my choice to be a law. When I make the original choice, I have no other reason for taking my end to be absolutely good, other than that it is good *for me*. This suggests that the pertinent fact about me is simply that I am the sort of being *for* whom things can be good or bad, that I am a being with interests.

Of course, someone might insist that I respect my own autonomy in a different sense—not in the sense that I treat a choice of my own as a law, but in the sense that I presuppose that what is good for autonomous rational beings, and only for autonomous rational beings, should be treated as good absolutely. But that conclusion is not driven by the argument, for there is no reason to think that because it is only autonomous rational beings who must make the normative presupposition, the normative presupposition is only *about* autonomous rational beings. Notice, too, that many of the things that I take to be good for me are not good for me merely insofar as I am an autonomous rational being. Food, sex, comfort, and freedom from pain and fear are all things that are good for me insofar as I am an animate being. So it is more natural to think that the presupposition behind rational choice is that the things that are good for beings for whom things can be good or bad are to be treated as good or bad absolutely. But, of course, things can be good or bad, in the relevant way, for any sensate being; that is, for any being who can like and dislike things, be happy, or suffer.[12] This suggests that the presupposition behind rational choice is that animals, considered as beings for whom things can be good or bad—as beings with interests—are ends in themselves.

We might put the point this way. As rational beings, we need to justify our actions, to think there are reasons for them. That requires us to suppose that some ends are worth pursuing, are absolutely good. Without metaphysical insight into a realm of intrinsic values, all we have to go on is that some things are certainly good or bad *for* us. That then is the starting point from which we build up our system of values—we take those things to be good or bad absolutely—and in doing that we are taking *ourselves* to be ends in ourselves. But we are not the only beings for whom things can be good or bad; the other animals are no different from us in that respect. So we should regard all animals as ends in themselves.[13]

9.3. WHY WE HAVE MORAL DUTIES *TO* ANIMALS

But there is another way to understand Kant's argument against the moral claims of animals. In a passage I quoted earlier, Kant says:

> As far as reason alone can judge, a human being has duties only to human beings
> (himself and others), since his duty to any subject is moral constraint by that
> subject's will. (MM 6:442)

One might place the emphasis here on the idea of owing a duty *to* someone, and take Kant to be claiming that it is impossible for us to owe a duty to an animal. It is, after all, notorious that Kant claimed that although we *do* have duties to treat animals humanely, we do not owe those duties to the animals, but rather to ourselves (MM 6:442; LE 27:459).[14] This claim goes right to the heart of the issue about legal rights for animals, since the duty of respecting a legal right is something that is supposed to be owed *to* the right holder. If we cannot owe duties to animals, then it seems that they cannot have rights.

In the passage I just quoted, Kant claims that to owe something to someone is to be constrained by his will. To see what this means, consider, first, what happens when you make a promise, and so incur an obligation. As Kant understood promises, what happens when you make a promise is that you transfer the right to make a certain decision, which is naturally your own right, to someone else, in rather the same way you might transfer a piece of property to someone else. If I promise to meet you for lunch at the cafeteria tomorrow, I transfer my right to decide whether to go to you, and

I now no longer have the right to decide that I will not go unless you absolve me from my promise. So my decision now belongs to you—it is a matter for your will to determine, not for mine. So you are in a position to constrain me to go to the cafeteria by your will. You can obligate me.

There is another way to understand this same transaction, which is again in terms of the making of a law. As we will see later, Kant envisions the original acquisition of a piece of property as the making of a kind of law that binds everyone. For example, when I claim a piece of land as my own, I in effect say that no one may use this land without my permission, and that everyone is bound by my will about how this land may be used. But Kant thinks that I cannot make laws for everyone else unilaterally, since other people are free and not bound by my will. So if I am able to make laws of this kind, to claim things for my own, it can only be by speaking in the name of what Rousseau called the General Will; that is, in the name of the laws we will together.[15] So when I make a promise, and so transfer my right to make a decision to you, we can understand that as our making a law together. When I promise to meet you and you accept my promise, we make a law together that my decision whether to meet you should belong to you and not to me. If our promises are mutual—if we promise to meet each other for lunch tomorrow—we both will the law that both of us should show up at the cafeteria tomorrow, and now neither of us can rescind the plan unilaterally. If I want to do something else, I have to get your permission, and if you want to do something else, you have to get mine. Having joined our wills under common law, we can only change things by making a new law together.

This gives us a way to understand those rights that are not incurred by particular actions, like the standing right not to be used as a mere means to someone else's ends. As we have seen, Kant supposes that all of us will that rational beings should be treated as ends in themselves, since (he thinks) the presupposition that rational beings should be treated as ends is built into every act of rational choice. So this is a law that, insofar as we are rational beings, we will together. The fact that we will it together is what makes it possible for us to make claims on each other in its name: we can bind one another through our wills. But the other animals do not participate in making moral laws, nor are they under the authority of those laws. They therefore cannot obligate us in the name of moral laws, and so cannot make moral claims on us.

So understood, Kant's argument is a version of what I call a "reciprocity argument." A reciprocity argument holds that human beings have either no duties at all, or no duties of justice (i.e. duties associated with rights), to the other animals, because such duties depend on relations of reciprocity. There are various versions of the argument. One is a crude picture of morality as a kind of social contract or bargain, whose content is something like: "I will act with a certain kind of restraint towards you, if you will act with a similar restraint towards me." This version prompts the obvious question of how we are to explain our duty to keep the social contract itself. That duty cannot be grounded in the contract.

Another version is associated with David Hume's argument that the requirements of justice only hold in certain conditions, conditions that John Rawls later called "the circumstances of justice" (Rawls 1971, sec. 22). Hume makes the argument in order to prove that the requirements of justice are grounded in considerations of utility. We expect people to conform to the requirements of justice only under certain conditions, he argues, and those conditions are exactly the ones in which conforming to the requirements of justice is useful to all concerned. Therefore it must be the utility that grounds the requirements. One of these conditions is an approximate equality of power between the parties to the social contract, which renders it in the interest of all parties to make and maintain the contract. On these grounds, Hume argues that we do not have duties of justice to the other animals:

> Were there a species of creatures intermingled with men, which, though rational, were possessed of such inferior strength, both of body and mind, that they were incapable of all resistance, and could never, upon the highest provocation, make us feel the effects of their resentment; the necessary consequence, I think, is that we should be bound by the laws of humanity to give gentle usage to these creatures, but should not, properly speaking, lie under any restraint of justice with regard to them ... Our intercourse with them could not be called society, which supposes a degree of equality; but absolute command on the one side, and servile obedience on the other. Whatever we covet, they must instantly resign: Our permission is the only tenure, by which they hold their possessions: Our compassion and kindness the only check, by which they curb our lawless will: And as no inconvenience ever results from the exercise of a power, so firmly established in nature, the restraints of justice and property, being totally useless, would never have place in so unequal a confederacy.
>
> This is plainly the situation of men, with regard to animals; and how far these may be said to possess reason, I leave it to others to determine. (Hume 1975, 190–91)

Hume's version of the argument seems subject to the objection that if some group of people acquired sufficient power over the rest of us, they would cease to owe us justice. Suppose, for example, that a small coterie of people obtains joint control over the only weapon capable of blowing up certain major cities, and uses the threat of doing so to blackmail the rest of us into submission to their will. Since it is not in their interest to cooperate with us, by Hume's argument, they are not obligated to act justly towards the rest of us. Hume seems even to invite that objection, for he emphasizes that in order to have the kind of superior power that frees people from the obligation to concede rights to others, it is not enough that the members of one group be stronger individually than the members of the other, they must also be sufficiently *organized* among themselves to maintain their force against the members of the weaker group:

> In many nations, the female sex are reduced to ... slavery, and are rendered incapable of all property, in opposition to their lordly masters. But though the males, when united, have in all countries bodily force sufficient to maintain this severe tyranny, yet such are the insinuation, address, and charms of their fair companions, that women are commonly able to break the confederacy, and share with the other sex in all the rights and privileges of society. (Hume 1974, 191)

I will come back to this point later, because it brings out something important about our relationship to the other animals. Meanwhile, notice that Kant's argument may be seen as a version of the reciprocity argument, for he thinks it is only those who stand in a certain kind of reciprocal relations with each other who can bind each other by law.[16]

If the reciprocity argument works, it captures something right about Kant's thought that the humane treatment of animals is something that we owe to ourselves. At least, insofar as the party *to whom* we owe a duty is the one who issues the law that gives us the duty, it is above all to ourselves that we owe it to treat the other animals humanely.[17] But the trouble with this thought, at least as far as *moral* obligation is concerned, is that Kant thinks that the ultimate foundation of moral obligation, *in general*, is autonomy, the rational being's capacity for issuing laws to himself. Even on Kant's own account, we are bound by the moral law because we ourselves will that rational beings should be treated in certain ways. Morally speaking, you have the capacity to obligate me through your will only because it is the law of my own will that I should respect your choices. Suppose my earlier argument is correct, and we ourselves are committed to the principle that all beings for whom things can be good or bad, all beings with interests, should be treated as ends in themselves. Then even if animals cannot obligate us through their wills, they can obligate us through their natures, as beings of that kind. For according to that argument, every act of our own will commits us to the view that such beings are ends in themselves, and as such are laws to us.

9.4. WHY ANIMALS SHOULD HAVE LEGAL RIGHTS

The argument I have just given, however, applies only to the *moral* claims of animals. In the case of legal or political rights, there is again an additional problem, closely tied to the problems I mentioned at the beginning of the paper. According to Kant, the sense in which others can obligate us legally is different from the sense in which they can obligate us morally (MM 6:218–21). The sense in which others can obligate us legally does not "go through" our own autonomy in the way I described above. Rather, the sense in which people can obligate us legally is that they may legitimately use coercion to enforce their rights. Coercion, as I mentioned at the beginning, may legitimately be used only for the sake of protecting freedom, a kind of freedom that the other animals, not being rational, apparently do not have. If the point of animal rights is simply to protect their interests, not to protect their freedom, then there seems to be no room for animal rights in a Kantian account.

But a closer examination of Kant's own argument again reveals grounds for questioning this conclusion. Earlier we saw how Kant grounds our claim to be ends in ourselves by showing that it is a presupposition of rational choice—a claim that is in a sense built into every act of rational choice. When I pursue the things that are good for me as if they were good absolutely, I commit myself to the principle that beings for whom things can be good or bad are ends in themselves. In much the same way, Kant tries to show that a commitment to enforceable rights for everyone, and therefore to a political state with a legal system, is built into every claim of right that I make for myself.

Here is how the argument goes: A legal or political right, as Kant understands it, is an authorization to use coercion. To say that you have a legal right to some piece of property is to say that if someone attempts to use it without your permission, you may legitimately use force to prevent him from doing so. But coercion is only legitimate when it is used in the service of freedom. Why then may we use it to defend our property? Like others in the social contract tradition, Kant envisions a state of nature in which people lay claim to parts of the commons for their own private use.[18] If it were not possible to claim objects as our own, Kant argues, we could not effectively use them when they were not in our physical possession. Or even if we could, our use of them would be subject to the will of others in a way that is inconsistent with our freedom. I cannot effectively grow wheat on my land if you might move in at any time and grow beans there, and I cannot do so freely if the only way I can do it is in effect to get your permission. In order to make free use of the land, I must be able to claim a right to it. A piece of property is a kind of extension of one's freedom. To deny the possibility of claiming objects in this way would amount to placing an arbitrary restriction on freedom (MM 6:246). Therefore we must concede that such claims—claims of enforceable right—are possible. Kant calls this "the postulate of practical reason with regard to rights" (MM 6:246).

So I can make it a law for you that you cannot use a certain piece of land without my permission. But I cannot do this unilaterally, since I am not your master. Rather, as we saw before, my claims of right must be made in the name of laws that have authority for us both, laws that we make together. In Rousseau's language, my claim must be made in the name of the General Will in order for it to have the force of law. Rights, Kant argues, are only "provisional" in the state of nature, since they cannot be fully realized until everyone's rights are protected by actual, coercively enforced laws, by a state with a legal system (MM 6:255–57). This is why it is a duty for us to leave the state of nature and live in political society. Kant calls this the "Postulate of Public Right" (MM 6:307). I am going to call Kant's two postulates, taken together, the "Presupposition of Enforceable Rights."

Since we must survive, we have to claim pieces of property for our own use, just as since we must act, we have to make rational choices to pursue certain ends. If the rational pursuit of my ends involves the presupposition that I have the right to use certain objects in pursuit of my ends, and that in turn involves the presupposition that everyone's rights should be upheld and enforced, then the Presupposition of Enforceable Rights is built into the rational pursuit of my ends. This exactly parallels the way that the presupposition that beings with interests should be treated as ends in themselves is built into the rational pursuit of my ends.

But who exactly is the "everyone" whose rights should be enforced? It is only rational beings who must lay claim to rights, and only rational beings who hold one another to the presuppositions of those claims, just as it is only rational beings who choose to pursue their ends and are rationally bound by the presuppositions of their choices. Before, we saw that it does not obviously follow that the presupposition behind rational choice is that *rational beings* are ends in themselves, and in fact when we looked more closely at the context in which the presupposition operates at the most basic level—namely, in my decision to pursue something simply because I think it will be good for me—it does not seem to follow at all.

Rather, what follows is that I am committed to the idea that if I am the sort of being for whom things can be good or bad, a being with interests, then I should be treated as an end in itself. In this case, too, we need to look more closely at the context in which the presupposition of enforceable rights first operates, which is the context of original acquisition.

But here we run into a problem. Although the problem is a general problem about ownership rights, it will be useful to pose it first as a problem about our rights (that is, the rights of human beings) to own animals. This will enable us to ask a question that we should be asking anyway, which is this: Even if it were not the case that the other animals could have rights against us, how exactly is it supposed to follow that we have rights over them? Putting the problem more generally, why is it supposed to follow from the fact that we need to claim objects as our own in order to use them effectively and freely, that we can claim anything we find in the world, even an animate being with a life of its own, that is not already claimed?

In the traditional doctrines of rights developed in the seventeenth and eighteenth centuries, especially in the theories of Locke and Kant, it is perfectly clear why this is supposed to follow. It follows from two theses. The first is a view originally derived from Genesis that found its way into these theories. That is the view that God gave the world and everything in it to humanity to hold in common.[19] The second is a picture of what a right in general is, a picture associated with the reciprocity argument. To claim that I have a right is to make a relational claim—and the relation is not between me and the object to which I have a right, but rather between me and other people. When we put these two claims together, we get a certain picture of what the general problem of individual rights is, a picture that is explicit in and familiar to us from the work of Locke, but also implicitly at work in the Kantian views we have just been reviewing. The problem of individual rights is conceived as a problem about what gives someone a right to take something out of the *commons*; or, to put it more carefully, about how I can take something out of the commons in a way that is justifiable to everyone else. Both Kant's insistence that rights must be established in accordance with the General Will and Locke's famous proviso—that the one who claims a right must leave as much and as good for others—are based in part on this picture.[20] Indeed, Kant insists on the essential role of this assumption in his theory. The "real definition" of a right to a thing, Kant says,

> is a right to the private use of a thing of which I am in (original or instituted) possession in common with all others. For this possession in common is the

only condition under which it is possible for me to exclude every other possessor from the private use of a thing [. . .], since, unless such a possession in common is assumed, it is inconceivable how I, who am not in possession of the thing, could still be wronged by others who are in possession of it and are using it. (MM 6:261)

Kant's assumption is slightly different from Locke's, because he distinguishes *possession* from *ownership*, properly speaking, and it is common possession that he posits. When something is in my physical possession, anyone (that is, anyone who is not its rightful owner) who tries to use it without my permission wrongs me, because she has to use force to get it away from me. This much follows simply from my innate right to freedom, which Kant understands to include control over my own body. When I *own* something, someone who uses it without my permission wrongs me, even when I am *not* in physical possession of it. The assumption of common possession seems less extravagant, for in a way it is simply the claim that no one has a prior right that would make it legitimate for them to exclude us from using the earth and its resources, and therefore to exclude us from dividing it up into property. Either way, however, the role of the assumption is to answer an obvious question: How could our agreement to divide the world up in a certain way have any authority, if we had no right to it in the first place?

Despite its religious formulation, the claim that God gave us the world in common captures an idea that goes right to the heart of the moral outlook, and that can be formulated in secular terms. It is the idea that others have just as good a claim on the resources of the world as we do, and that it behooves us to limit our own claims with that in mind. But the idea of the world as owned or possessed in common *by humanity* also represents the world, and everything in it, including the animals, as one big piece of property. That Kant was prepared to represent the world in this way is important, because it shows that Kant had no principled reason for regarding animals as possible property. He simply assumed that that is what they are.

At the beginning of this paper, I said that it is inconsistent with Kant's methodology simply to accept metaphysical claims about value. Claims about value, like any claims that go beyond the realm of empirical experience, must be established in a certain way. They must be shown to be necessary presuppositions of rational activity. The claim that world is *given to us* in common is certainly such a claim, for it is not scientifically provable. Is it just a religious or metaphysical claim that really should have no place in Kant's philosophy? Or could we regard it instead as a presupposition of rational activity? In fact, in its modified form as the presupposition of common possession, Kant explicitly claims that we can:

> All human beings are originally (i.e. prior to any act of choice that establishes a right) in a possession of land that is in conformity with right, that is, they have a right to be wherever nature or chance (apart from their will) has placed them. This kind of possession . . . is a possession *in common* because the spherical surface of the earth unites all places on its surface. . . . The possession by all human beings on the earth which precedes any act of theirs that would establish rights . . . is an *original possession in common* . . . , the concept of which is not empirical. . . . Original possession is, rather, a practical rational concept which contains a priori the principle in accordance with which alone people can use a place on the earth in accordance with principles of right. (MM 6:262)

Before there are any other rights, before we start dividing up the world for our purposes, each of us has a right to be where he or she is, wherever "nature or chance" has placed us.[21] The right to be where you are is an aspect of your right to control over your own body, since it means that in the absence of prior claims, no one has a right to force you to move on. Since a right to the earth, for Kant, goes with a right to use its resources for your support, that means that each of us has a right to take what he or she needs in order to live.

In other words, we are thrown into the world, and having no choice but to use the land and its resources in order to support and maintain ourselves, we have no choice but to assume at least that we are doing nothing wrong in doing that. But we are not the only creatures thus thrown into the world, with no choice but to use the earth and its resources in order to live. If this is the basis of the presumption of common possession or ownership, why not assume that the earth and its resources are possessed in common by all of the animals?[22]

Again, it is true that rational beings are the only animals who must conceive of their situation in these normative and moral terms, and therefore the only beings who must presuppose that we have a right to use the earth for our maintenance. But it does not follow that *what* we have to presuppose is that rational beings, and rational beings alone, have that right. In the absence of a prior religious commitment, it is arbitrary to make any assumption except the assumption that the world belongs in common to all of the creatures who depend on its resources. Only some sort of metaphysical insight into a special relationship that human beings stand in to the universe could justify the assumption that it belongs only to us, and that is exactly the sort of metaphysical insight that Kant denies that we have. To the extent that the kind of "freedom" that is at stake in rights is simply the freedom to use your own body to carve out some sort of a decent life in the world where you find yourself, then the "freedom" of the other animals is the sort of thing that could be protected by rights after all. [23]

Of course, despite what Kant plausibly says about its necessity, we could drop the presupposition of common possession or common ownership altogether. But if we do that, we must also drop the version of it that comes down to us from Genesis. In that case, the world was not given to human beings in common, because it was not given to anyone. That means that what human beings have over the other animals is not, in general, a form of rightful ownership. *It is simply power.*

Now recall that the starting point for Kant's theory of why we must conceive ourselves to have rights is the wrongness of the unilateral domination of some individuals by others. It is the wrongness, to put it more colloquially, of the view that might makes right. The reason why the political state and its legal apparatus exists at all, according to Kant, is not that fighting over everything all the time is inconvenient, or that life in the state of nature is, as Hobbes famously reminded us, nasty, brutish, and short. It is the urgency of standing in relations with others that we can regard as rightful that prompts us to establish a system of enforceable legal rights.

But human beings, collectively speaking, do stand in relations of unilateral domination over the other animals. I am not talking now about a relation in which we as a species stand to them as species. I am talking about a relation in which human beings stand as an organized body to individual animals who are not part of any such body. To us, the other animals are a subject population, rendered almost completely at our mercy by our intelligence, power, and organizational skills.[24]

In fact, when Hume describes the relations in which people stand to animals, he is describing *exactly* the sort of unilateral domination of some beings by others the wrongness of which is the starting point of Kant's political philosophy. And when he talks about the relations in which men stand to women, Hume, with his characteristic political realism, brings out the important thing that makes such unilateral domination possible: that the dominant group be able to organize itself as a group, while members of the dominated group can only resist as individuals, if indeed they can resist at all. This is an essential feature of the relationship in which human beings stand to the other animals. And the way that we unite and organize ourselves is by constructing our legal systems.

Earlier I pointed out that the problem Kant has in mind when he constructs his account of rights does not concern the probability of bad behavior. He thinks it is wrong in itself for one person to be completely subject to another person's will. Unilateral domination is a moral wrong whether it is abused or not. But I did not say that unilateral domination is not the *source* of bad behavior—and notoriously, it is. You need only look at what goes on inside of our factory farms and experimental laboratories to see what the possibility of such domination—the ability to do whatever we like with another animal—can lead to. So long as there are profits to be made, and the tantalizing prospect of expanding the human life span by experiments on the other animals, there will be people who will do *anything*, no matter how cruel it is, to a captive animal. And what makes this possible is the legal status of animals as property. It is not plausible to hope that the human race will someday have a collective humanitarian conversion and bring all such practices to an end, without any help from the law. But even if it were, the argument would stand. No matter how well-intentioned we are, we can only be rightly related to our fellow creatures if we offer them some legal protections.

If we must presuppose that the world and all that is in it is possessed by us in common, so that we may use it rightfully, then we should presuppose that it is possessed by all of its creatures on the same ground. The other animals are not part of what we own, to use as we please, but rather are among those to whom the world and its resources belong. If we reject the presupposition of common possession or ownership, then we cannot pretend that the way we treat the other animals is anything but an exercise of arbitrary power, the power of the organized over the weak. In that case, I suppose it is up to us how we treat them—but the moral argument still holds. The other animals are, just as much as we are, beings with interests, beings for whom things can be good or bad, and as such they are ends in themselves. Either way, the only way we can be rightly related to them is to grant them some rights.

9.5. CONCLUSION

Despite his own views about animals and their claims, Kant's philosophy captures something about our own existential situation that proclaims our fellowship with the other animals. It is the central insight of Kant's philosophy that the laws of reason are our laws, human laws, and that we cannot know whether the world as it is in itself conforms to them or not. The fact that we are rational does not represent a privileged relationship in which we stand to the universe. Kant also believed that morality is a kind of substitute for metaphysics, giving us grounds to hope for what we cannot know through any metaphysical insight—that the world can, through our efforts, be made into a place that meets our standards, that is rational and good.[25] This means that we share a fate with the other animals, for, like them, we are thrown into a world that gives no guarantees, and that are faced with the task of trying to make a home here. It is a presupposition of our own rational agency and of our moral and legal systems that the fate of every such creature, every creature for whom life in this world can be good or bad, is something that matters. That is why we should concede the moral claims of the other animals, and protect those claims as a matter of legal right.

ACKNOWLEDGMENT

An earlier version of many of the arguments in this paper was discussed in a workshop on my work on ethics and animals at New York University in September 2011 and by an ethics reading group at Stanford in October 2011. I would like to thank Charlotte Brown, Tom Doherty, Lori Gruen, Beatrice Longuenesse, Peter Singer, and Jeremy Waldron for comments.

NOTES

1. Kant's works are cited in the traditional way, by the volume and page number of the standard German edition, *Kants Gesammelte Schriften*, edited by the Royal Prussian (later German) Academy of Sciences (Berlin: George Reimer, later Walter de Gruyter & Co., 1900–), which are found in the margins of most translations. The abbreviations I have used are as follows; for the translations used, please see the bibliography.
   C3 = *Critique of Judgment*
   C2 = *Critique of Practical Reason*
   CBHH = "Conjectures on the Beginnings of Human History"
   G = *Groundwork of the Metaphysics of Morals*
   LE = *Lectures on Ethics*
   MM = *The Metaphysics of Morals*
2. I have changed Nisbet's rendering of the German *Pelz* from "fleece" to "pelt" although the German can be rendered either way, because I think that the rendering "fleece" softens Kant's harsh point.
3. Rawls 1971; first used on xii.
4. The contrast here is with Locke and Hobbes, who supposed that we leave the "state of nature" as a remedy for its "inconveniences" (the word is Locke's) and therefore from motives of prudence rather than because it is morally required. See Locke 1980 and Hobbes 1994.
5. This rough description of Kant's method skates over a great many complexities and controversies in Kant interpretation. What I am calling "presuppositions" are of various kinds—constitutive principles, regulative principles, and postulates,

for instance; and the arguments Kant gives to validate them are also of various kinds—the special kind of argument he calls "deduction," for one; in the case of the argument for the moral law in the Second *Critique*, the establishment of a "credential" (C2 5:48), for another; and others as well. In addition, there is philosophical controversy over the nature of the specific validation Kant ultimately proposed for the moral law, and Kant himself changed his mind about this over the course of his career. Despite these complications, I think that the rough description of Kant's method generally fits all these cases.

In this paper my focus is on the presuppositions themselves, not on their validation. I will argue that in certain ways Kant misidentified the presuppositions of practically rational activity. That leaves it open, I suppose, that the revised presuppositions cannot be validated. Because of the great obscurity of Kant's methods of validation, especially in moral philosophy, it is a little difficult for me to address this worry in general terms, but I do not believe this is a problem. I think Kant was right in concluding that the presuppositions of rational action do not need a deduction in the same sense that the presuppositions of theoretical understanding do.

6. I first presented a version of this interpretation of Kant's argument in "Kant's Formula of Humanity" (Korsgaard 1996a).

7. This is not to say that weakness of the will and moral weakness are impossible, of course, but it implies that they must be explained in terms of self-deception.

8. Someone might, of course, challenge Kant's claim that the adequacy of one's reason implies that one's end is an end for everyone. Kant's assumption is that reasons are what I have elsewhere called "public," or what are sometimes called "agent-neutral," reasons—reasons whose normative force extends to all rational beings. I have defended this assumption in various places, including *The Sources of Normativity* (Korsgaard 1996c, Lecture 4) and *Self-Constitution* (Korsgaard 2009, chap. 9). It would take me too far afield to discuss this complex issue here. I assume that the primary audience for the argument of this paper is people who are prepared to grant that human beings or rational beings have legitimate moral and legal claims on each other, and who therefore are prepared to grant that in some sense we are laws to each other, even if they are unsure whether the other animals also have such claims on us.

9. I say, "or at most" because of course we might desire things that are bad for us, that are inconsistent with our happiness, and that are not rational to choose on that ground.

10. The point of the caveat in the first clause is to leave room for duties owed to God, and grounded in faith. Since Kant thinks we cannot prove there is a God who is a rational being with a will, or have theoretical knowledge of what God's will is, we cannot owe duties to God "as far as reason alone can judge." This is not inconsistent with Kant's occasional suggestion that we should view God as the sovereign of the Kingdom of Ends (G 4:433, 4:439). There is a sense in which Kant himself thinks faith itself is grounded in reason, but it is not the usual sense: Kant does not think that there are successful theoretical arguments for the existence of God and the possibility of a future life. Rather, he thinks our moral commitments require us to hope that a fully moral state of the world can be achieved, and that the "postulates" of God and Immortality, the objects of "practical faith," give us a picture of the conditions under which a morally perfect world could be achieved. Sadly, Kant did not envision that morally perfect world as including eternal happiness for the other animals. Rather, he says that without such faith, all that

even the best person can expect is "deprivation, disease, and untimely death, just like all the other animals of the earth" (C3 5:452). For further reflections on this aspect of Kant's moral philosophy, see "Just Like All the Other Animals of the Earth" (Korsgaard 2008).

11. See *Self-Constitution* (Korsgaard 2009, sec. 4.5) for a fuller version of this argument.

12. There is a sense in which things can be good or bad for any functionally organized being—namely, things can help or hinder its functioning. "Riding the brakes is bad for your car," we say in that sense. The car, however, is made for a human purpose, and the way in which things can be good or bad for it is derivative from that purpose: ultimately, what happens to the car is good or bad for people, not really *for* the car. Things can also be good or bad for plants, and this kind of goodness and badness is not derivative from human purposes ("The weeds are really flourishing in my garden; all this rain is good for them."). Rather, it is good for the plant considered as a living organism, functioning so as to survive and reproduce. The way in which things are good or bad for people and animals includes this, but adds a new dimension, for an animal has a point of view on which the things that are good or bad for it have an impact—they are also good or bad *from* the animal's point of view. In saying this, I am not endorsing the hedonistic conclusion that only experiences themselves can be good or bad, insofar as they are pleasant or painful. I am only suggesting that there is a sense of "good for" in which "good for" and "bad for" are relative to the evaluative attitudes of the being for whom things are good or bad. By "evaluative attitudes" I mean desires, pains, pleasures, fears, loves, hates, ambitions, projects, and principles, and so on, some of which are experienced by every sensate being. This is the sense of "good for" that I take to be relevant to the argument. For further reflections, see "The Origin of The Good and Our Animal Nature" (Korsgaard forthcoming).

13. The main argument of this section was first advanced in "Fellow Creatures: Kantian Ethics and Our Duties to Animals" (Korsgaard 2005).

14. In fact, Kant's views were rather advanced for his day. Kant thought animals should not be hurt or killed unnecessarily, and certainly not for sport (LE 27:460). If they must be killed, it should be quickly and without pain (MM 6:443). We should never perform painful experiments on them for merely speculative purposes, or if there is any other way to achieve the purpose of the experiment (MM 6:443). We should not require harder work of them than we would require of ourselves (MM 6:443). When they do work for us we should we treat them as members of the household (MM 6:443), and when they no longer can work for us, they are entitled to a comfortable retirement at our expense (LE 27:459). Nonhuman animals, according to Kant, are the proper objects of love, gratitude, and compassion, and failing to treat animals in accordance with these attitudes is "demeaning to ourselves" (MM 6:443; LE 27:710).

15. Rousseau 1987. The term is first used in chapter VII, p. 26.

16. For a more detailed account of Kant's argument as a reciprocity argument, see "Interacting with Animals: A Kantian Account" (Korsgaard 2011a). Notice, however, that Kant's version of the argument does not fall prey to the objection I have just made to Hume. In Kant's argument, it is everyone's freedom, not everyone's interest, which is at stake, and you cannot legitimately claim a right without upholding everyone else's freedom. So the coterie of powerful people would still owe the rest of us justice.

17. Kant also sometimes suggests that the reason we owe humane treatment to the other animals is that our treatment of other human beings is likely to be influenced by our treatment of the animals (MM 6:443; LE 27:459). Although it is now notorious that there is a connection between serious criminal behavior and animal abuse, the suggestion is a peculiar one for Kant to make. After all, if reason really did tell us the animal suffering does not matter in the way that human suffering does, why would we be tempted to treat humans in the same way we treat animals?

18. I will come back to the role of the idea of the commons in these arguments below.

19. Actually, in Genesis 1:29–30, God gives the plants to human beings, and then in Genesis 9:3, God gives everything living and moving to human beings.

20. . V, paragraph 33, p. 21).

21. Readers of Dickens (1977) may remember the poor boy Jo, who does *not* have a right to be where he is—he is always being told by the constable to "move on"—and consequently has no rights at all.

22. As I mentioned in note 19, in the Genesis account the plants of the world are given to the animals before the animals are given to human beings.

23. Of course I am not suggesting that the correct way to protect the lives of animals is arrange for them to them own property. Nor is there any hope of dividing up the world in a way that leaves "enough and as good" for every animate creature when some of them must live by preying on others. But in the case of wildlife, we might think some duties of habitat preservation do follow, and domestic animals certainly have a right not to be starved. The most suggestive thought here is that if animals do have this kind of freedom, we do not have a right to their bodies—those are not ours to do with as we please. But the details of what is required by the arguments of this paper, of which rights animals should have, remain to be worked out.

24. Here it is important to remember that I am talking about how human beings stand collectively to animals individually. As species, many of them are also subject to our domination, and it is up to us whether many species will survive. But that certainly is not true of all of them. Collectively speaking, the mosquitoes may defeat us yet.

25. See note 10.

10

Kantian Constructivism and the Ethics of Killing Animals

FREDERIKE KALDEWAIJ

10.1. INTRODUCTION

This paper considers the ethics of killing animals from the perspective of "Kantian constructivism." This label covers a group of contemporary moral theories with a specific view of the nature and basis of moral obligation. These theories are *constructivist* in that they rely on a procedure to determine which reasons are valid, where this procedure is justified by showing that it is implicit in the practical point of view of an agent (or agents). This is in contrast to realist theories, which hold that the validity of reasons is independent of what agents contingently or necessarily accept. These theories are *Kantian*, in that this procedure involves, roughly, the (possibility of) agreement of all concerned.

Traditionally, Kantian theorists do not count nonhuman animals among those who must be able to agree with our reasons for action, perhaps mostly because they are not thought to possess the capacity to agree in a relevant sense. Following Kant, they regard the capacity for rational deliberation (and sometimes specifically moral autonomy) as a necessary condition for having moral status. They do accept that we have duties *regarding* animals, but these are seen as *indirect* duties, ultimately owed to rational beings.[1] Whether duties regarding animals are direct or indirect makes both a symbolic and practical difference. Symbolically, if we merely have indirect duties regarding animals, we will not treat animals in certain morally required ways *for the animals' sake*. Practically, indirect duties are comparably less strict than direct duties, as such duties do not involve *owing* certain treatment to animals (in a way that is correlative with animal rights).[2]

Some have argued against the rejection of direct duties to animals on grounds *external* to Kantianism, for instance by referring to our considered intuitions about our duties to animals, or to human beings with comparable capacities (see e.g. Broadie and Pybus 1974; Regan 2004). In this paper, I will argue against this claim from *within* contemporary Kantian constructivism. I have two reasons for this. First, Kantianism is one of the major *normative* moral theories of our time. Many endorse the Kantian ideal of acting only in ways that others can agree with. It is therefore interesting to determine whether there are good internal reasons for Kantians to accept duties to animals, and to see what this implies for the ethics of killing them.

The second, and most important, reason is *metaethical* in nature. It is worth critically investigating whether the Kantian constructivist method, which understands moral duty in terms of what rational agents have to *accept*, implies that only duties *to* rational beings are possible. Constructivism is now a hotly debated method in ethics (see e.g. Lenman and Shemmer 2012; Bagnoli 2013). This is especially true for the more ambitious versions. "Global" or "thoroughgoing" forms of Kantian constructivism claim that commitment to the Kantian ideal of acting only in ways that others can accept is not conditional on the moral values of the agent; instead, *any* action or interaction rationally commits one to this ideal. Such arguments offer the promise of a fundamental justification of moral duties that is metaphysically very modest.

In what follows, I shall first explain Kantian constructivism in more detail. Then I shall argue that the ambitious Kantian constructivist arguments, if they justify substantive duties to others at all (including rational beings), also justify duties to certain nonhuman animals. Next, I will discuss what the normative implications of this argument are with regard to the ethics of killing such animals. I will discuss three reasons why killing animals can be thought to be problematic in a Kantian constructivist framework, two of which link up with discussions about the harm of death in the first part of this volume. Finally, I will briefly point out some notable practical implications regarding the ethics of killing animals. Unlike utilitarianism (the other influential moral theory that is discussed in this volume), these arguments imply a *strict* duty not to kill (unless this is the only way to avoid violating an overriding duty).

10.2. KANTIAN CONSTRUCTIVISM

As mentioned above, the label "Kantian constructivism" is here used for a group of theories that (roughly) regard reasons as valid when everyone involved can agree with them (under specific conditions).[3] While contemporary Kantian constructivists are all inspired by Immanuel Kant in some way, Kant scholarship is not their only or primary aim. In this paper, I am interested in the structure of contemporary "Kantian" arguments, and not in whether or not they are true to Kant. Let me first explain what constructivism is and describe some different forms of Kantian constructivism. Constructivism is a theory of the nature and justification of reasons. Constructivists take the validity of reasons to be determined by certain procedures or principles. These procedures, in turn, are justified by showing that they are constitutive of, or rationally implied by, something that an agent either contingently or necessarily accepts.

It is becoming commonplace to distinguish *local* or *restricted* constructivism, on the one hand, from *global* or *metaethical* constructivism, on the other (Street 2010; Lenman and Shemmer 2012). In local constructivism, the procedure that determines the validity of reasons is taken to be constitutive of specific, substantive normative starting points. For example, in the case of John Rawls's (1971) local Kantian constructivism, the "original position" in which principles of justice are chosen from behind the "veil of ignorance" (shielding people from knowledge of their gender, occupation, etc.), embodies the normative judgments that people are free and equal and society is a fair system of cooperation over time (Street 2010, 368). Restricted constructivism involves the claim that *if* agents accept these normative judgments, they can use this procedure to determine which reasons are valid, but it is not argued that *all rational agents* necessarily have to do so.

Global constructivist arguments, on the other hand, do not start from specific substantive normative views, but involve the claim that certain procedures are constitutive of acting for *any reason whatsoever*. Specific substantive reasons are valid when they withstand scrutiny from, or are endorsed on the basis of, these procedures.[4] While local constructivism is compatible with metaethical realism (because the assumed normative judgments could be held to be valid independently of the practical point of view of an agent), this is not the case for global constructivism (Street 2010). The validity of normative claims is understood entirely in terms of what agents have to *take* to be valid. Global *Kantian* constructivists argue that *moral* duties to others follow from the practical point of view of an agent. Valid reasons are not ultimately contingent on the particular values that an agent accepts (as in Humean constructivism), since everyone else has to be able to share the reasons, too.[5] Global Kantian constructivism, then, involves an ambitious attempt to justify moral duties without relying on any, possibly controversial, moral intuitions.

I shall call contemporary attempts to argue that morality is implicit in agency "arguments from agency." Christine Korsgaard (1996d, 2009) is the most prominent contemporary example of someone who presents such an argument.[6] Korsgaard argues that acting on the categorical imperative is a necessary condition for the possibility of rational agency. The categorical imperative comes in different formulas, one of which is a universality requirement: you must be able to will your reasons as laws for any rational agent.[7] The idea is that you cannot understand yourself to be a rational agent unless you take your reasons to be valid, not just for you as you are right now, but for anyone (including for your future self, which may have other desires than your current self) (Korsgaard 2009). Arguments from agency are held to establish that agents, insofar as they are rational, have to respect all other agents' purposes, or the generic conditions for successful agency (e.g. freedom and well-being).[8]

Other Kant-inspired contemporary authors are skeptical about the claim that moral obligation is implied by individual agency. They argue that it is instead implied by specific kinds of *inter*action: making claims on one another, or deliberating together.[9] Such "arguments from interaction" (as I shall call them) are also known as "second personal" (vs. "first personal") arguments (Darwall 2006). Prominent examples are Steven Darwall (2006), Onora O'Neill (1989) and, in the German-speaking realm, discourse ethicists like Karl-Otto Apel (1980) and Jürgen Habermas (1993). Interestingly, and similarly to arguments from agency, these arguments do not start with *specific* normative assumptions (e.g. that it is wrong for others to step on your feet).[10] The idea is that if we hold someone responsible for deliberating or acting in a certain way (whichever way this is), we implicitly assume that he has to be able to, on reflection, agree that this is indeed how he should act, and guide his actions in accordance. Two levels of norms follow from this kind of argument. First, there are norms involving the conditions that make deliberation *possible*, such as the norm not to use coercion and manipulation. Second, there are norms that are actually agreed upon in shared deliberation of the agents involved. This procedure can also be understood in terms of a universality requirement, but not in terms of what *I*, as an agent, can will as a universal law, but as what *we together* can will as a universal law. Some of these authors argue that we cannot rationally avoid deliberating together about reasons, while others hold that this is rationally avoidable, but not psychologically so, since we cannot help holding others accountable and being held accountable by others.[11]

In this paper, I will not defend the Kantian constructivist view that reasons are only valid if all concerned can accept them. Arguments from agency and interaction are ambitious attempts to provide sufficient justification for this normative view, but they are controversial (e.g. Enoch 2006; Street 2012). It may be thought impossible or unnecessary to justify moral duties "all the way down." But if we, like Rawls, already start with assuming that "people" are free and equal and have a concern for justice, this begs central questions that are interesting from the perspective of animal ethics. Why exactly is moral or rational agency thought to be necessary for moral status, and is this claim justified? As arguments from agency and interaction purport to answer these questions, it is interesting to examine them further.

10.3. THE MORAL STATUS OF ANIMALS

Traditionally, it is thought that the Kantian constructivist view of the *source* of our moral duties has implications for the *scope* of our moral duties, or whom we have duties *to*. Autonomy is a central concept in Kantian thought: the morally good agent acts on laws that are not imposed by an external authority, such as God, nature, or a tyrant, but (co-)legislated by the agent herself. This is in line with the central anti-realist ambition of global constructivism: the rational agent is the source of normativity. It is usually thought that this implies that we have duties to all and only agents who can "legislate laws" and act in accordance with them. In other words, we only have duties to those who can *deliberate rationally* about how they should act, by testing their reasons with the procedures that are constitutive of agency or interaction. I shall argue, however, that it does not follow from the Kantian constructivist method that we have to accept duties to only rational beings. If these arguments justify substantive duties to rational agents, they also justify duties to nonrational but purposive agents, which may include certain nonhuman animals. I will here only be able to offer a general sketch of the argument, and will not be able to go into the details of the various different versions of Kantian arguments from action and interaction. This section is a summary of arguments that I have elaborated elsewhere.[12]

While my argument goes against the grain of much of Kantian thought, I should note that I am not alone in arguing that Kantian constructivism implies duties to animals. A prominent Kantian philosopher is on my side: Christine Korsgaard. Korsgaard (1996a) originally argued that we must regard rational choice as the source of value and, therefore, respect all those who have the capacity of rational choice. She now argues that rational beings, by choosing to act for aspects of their natural good, confer normative value on the good of animals, which must be understood as "ends in themselves."[13] While I agree with Korsgaard on various points, my argument also differs from hers, mostly in that I claim that we cannot avoid accepting duties to animals, as they share the ultimate *basis* for our actions and interactions.[14] In chapter 9 of this volume, Korsgaard considers the issue of animal rights from the perspective of Kant's legal philosophy, while in this paper I will consider the question of what treatment we owe animals from various contemporary Kantian arguments for moral duties to others.

Let me briefly outline my argument, before going into the details. To determine *to whom* we have duties, I suggest that we determine the necessary and sufficient *basis* of the laws that agents legislate. If the *sufficient ground* of some moral duties is shared by other individuals, then moral agents have to accept duties to these individuals. I shall argue that the basis of moral duties, in these arguments, is a capacity we can plausibly be thought to share with certain nonhuman animals; namely, a type of purposive agency. While the capacity of rationality (meaning the capacity to examine one's reasons by light of the procedures that are constitutive of agency or interaction) is necessary to be morally *responsible*, it is not necessary to have moral *status*. The requirement of universalizability refers to whom we must take to *act* on the laws (all beings who *can* act on laws), not whom the laws *protect* for their own sakes.

I will now present the argument in more detail, first with regards to arguments from agency. These involve the attempt to show that moral duties are implicit in acting for any kind of purpose. *Agency* is therefore thought to be the source, or ultimate ground, of reasons and moral duties. But what *kind* of agency is this? Traditionally, it is thought that this is *rational* agency. But this cannot be so, for Kantian constructivists understand rational agency in terms of acting on the principles that are *constitutive of agency*. The claim that rational agency involves acting on the principles constitutive of rational agency is viciously circular. To put the same point in another way: the test of what we can "will" as a universal law cannot concern what we can "rationally will" or "morally will" as a universal law, since it *determines* what we can rationally or morally will. So we need some account of *prerational agency*, or "willing," to find out what we rationally ought to will.

How can we understand this idea of prerational agency? Recall that Kantian (global) constructivists are anti-realists. All value is conferred or projected by valuing creatures, and motivates agents to act. As we cannot presuppose *rational* motivations, it makes sense to look to what Thomas Nagel calls "unmotivated desires," or those desires we simply find ourselves having (1978, 29–30). Such desires involve an evaluatively loaded representation of activities or objects, such as a positive evaluation of playing and a negative evaluation of being attacked (see Korsgaard 2009). These desires are rooted in our physical and psychological make-up, and they may vary with our cultural and environmental circumstances. Note that nonhuman animals may well be thought to have desires of this type.[15] Of course, there are many discussions on animal thinking and consciousness in philosophy of mind and agency, but we commonly believe that certain animals act in a way mediated by something like nonlinguistic beliefs and desires (see Bermúdez 2007 for a philosophical defense of this view).

What nonrational animals presumably *cannot* do is take a step back from their immediate motivations and question whether they provide the agent with a *sufficient* reason to act (see Korsgaard 1996d). It seems likely that one requires language to be able to have higher-order thoughts about one's motives, and to reason formally (see e.g. Bermúdez 2007). Animals may reason in an informal way, understanding that they need to walk to their food bowl to be able to eat. *Rational* animals, like us, can formally understand that taking the means is constitutive of pursuing a purpose.[16] More importantly, we can also consider the ultimate source or basis of our purposes. Whatever specific purpose an agent pursues, if the agent acts on it *because* he or she has an unmotivated desire to fulfill this purpose, and she considers this to be a *sufficient reason* for action, this then commits her to moral conclusions (see also Korsgaard 2011b). The agent must consider prerational agency as sufficient for conclusive reasons for action.

One might not be convinced that the prerational agency that is the basis of these arguments must be the capacity to act on unmotivated desires that we share with certain nonhuman animals. What sets humans apart, it may be thought, is the capacity for *free* but for *free* choice. Recall that "free choice" in this context cannot be understood as based on reason (as the argument precisely has to establish what is rational) nor as based on an insight into the external value of reasons (realism). If free choice is not understood as choice on the basis of unmotivated desires or values, the only possibility that seems left is regarding it as a basic "going for something," in a voluntarist or existentialist way. Human beings are then thought to generally choose things they (or others) desire or find important, but not to do so, ultimately, *because* they (or others) desire them, but completely freely, "out of the blue." First of all, this seems like an ad hoc claim that cannot be proved. Second, far from explaining how rational action is possible, this makes "rational" action ultimately random and meaningless.

It makes far more sense to understand practical rationality, from a constructivist standpoint, in terms of acting in accordance with what is implicit in your prerational agency: making sure that there is no contradiction between your subjective purposes, the means you take to pursue them, and the reasons of others. The agent who considers unmotivated desires as a sufficient basis for actions is rationally required to respect the unmotivated desires (or evaluative dispositions) of others, too. She cannot accept any reasons for action that involve undermining a prerational agent's control of her behavior (e.g. through coercion or manipulation) or take away the necessary means to her successful agency (e.g. through physical harm) for no good reason.[17]

So far, the sketch is of the case for duties to animals as it applies to "arguments from agency." As explained above, proponents of arguments from *interaction* are skeptical about deriving duties to others from the perspective of an individual agent pursuing purposes. Can an individual agent not just, for example, take his reasons to be valid for anyone *who also has his purposes*, rather than for *anyone at all*? Rather, specific types of interaction, such as holding each other responsible, or engaging in a debate on the right thing to do, imply that we must leave each other free to deliberate, and to act in accordance with the conclusions of our deliberation. It seems that this only provides a basis for duties to rational beings, who *can* determine how to act—we must justify our actions to them in terms they can accept.

However, these arguments, too, imply duties to certain nonhuman animals, for similar reasons as arguments from agency. As we have seen, there are two levels of norms on the basis of such arguments. The first level involves requirements not to engage in actions that would undermine the very *possibility* of holding others responsible, or reasoning together, such as coercion or manipulation. These norms are merely *procedural*; they determine the conditions for shared deliberation. Then, on the second level, we actually have to deliberate together to determine what are valid *substantive* reasons for action. Various proposals may be given for reasons, such as "we should never help anyone in need when this is not in our interest." Note that these proposals for reasons cannot yet be taken to be *rational* proposals, since whether they are rational (or valid) is to be determined by the shared deliberative process. These proposals are based on prerational values or reasons. Note also that the first-level procedural rules are merely meant to make sure that nothing *distorts* the deliberative process, such as untruthfulness or power inequalities. They cannot determine which substantive norms are ultimately valid.

Kantian constructivists do not assume that the deliberative process should track moral facts that are *external* to the deliberation. The moral facts are determined by the deliberation *itself*, and the deliberation must be based on the participants' prerational, subjective values. To be able to come to intersubjective agreement, we have to consider everyone's input into the deliberative process on its own merits. For example, I may demand that someone get off my foot, since he is hurting me, and this itself may be regarded as a prima facie valid demand.[18] Whether this demand will ultimately be met depends on the other relevant subjective values (e.g. on whether stepping on my foot is not necessary to save someone from a bad fall). Even if the ultimate basis for shared reasons is subjective value, that does not leave us without things we can meaningfully argue, since there is still the significant challenge of determining what exactly everyone really values, and how we are going to be able to respect everyone's values, in the face of human nature, natural scarcity, the complexities of our social practices, and other empirical facts.

As we, rational beings, make claims on each other on the basis of prerational, subjective values, there seems to be no legitimate reason to exclude the similar values of animals from being taken seriously. Does this justify duties that we strictly *owe to* animals, though?[19] Does this not make animals like beautiful paintings, that are at most objects of our moral duties? Nonhuman animals are not among those who can deliberate with us and determine which moral duties we have to accept. (Neither are small children or intellectually handicapped adults). But we can distinguish between, on the one hand, the capacities of formal reasoning that allow us to determine and act in accordance with the constitutive principles of (inter)action, and, on the other, the basis on which we choose or make substantive claims in the first place: our unmotivated desires or values. Animals have a characteristic or capacity that we regard as a *sufficient basis for a valid claim* on others, even if they cannot actively make such a claim. Nonhuman animals should therefore be included in deliberation by means of advocates, who represent them as best they can. Most adult human beings, on the other hand, including those that have been previously denied access to "legislative authority" (such as women, or people of color), can reason prudentially and morally, and therefore there is no reason to exclude them from being among the "legislators."[20]

This does not yet amount to a full vindication of duties to animals on the basis of Kantian constructivism, since I have assumed the basic validity of these arguments. Do we *have* to share our reasons with everyone else for them to be valid? Are we really irrational if we only take the reasons into account of those we love or hold in personal esteem? These are interesting and important questions, but they are as relevant for the justification of duties to rational beings as they are for the justification of duties to animals. People who are skeptical of these ambitious arguments may prefer to start moral enquiries with presuppositions that (Western, liberal) people generally agree with, as Rawls does. Even if we do not accept the ambitious constructivist arguments, my reasoning should at least create an awareness of the distinction between being able to deliberate explicitly using formal procedures, on the one hand (say, choosing principles of justice from behind the "veil of ignorance"), and making choices on the basis of values (or acting for a "conception of the good"), on the other. We eventually act and interact on the basis of values, which are not given by reason alone, and do not come from nowhere, but are based in an evaluative nature that we seem to share with certain nonhuman animals.

10.4. IMPLICATIONS REGARDING THE KILLING OF ANIMALS

10.4.1. Introduction

I have argued that Kantian constructivist arguments for duties to others, if valid, also imply duties to certain nonhuman animals; namely, those that desire to fulfill certain purposes. What does this mean with regard to the ethics of killing animals? What interests me here is whether there is a basic duty not to kill, quite apart from a duty not to make someone suffer (for example, due to pain or stress caused by the act of killing). As we have seen, Kantian constructivist authors understand reasons in terms of what agents can accept, or what reasons they can share. However, a well-known philosophical puzzle about the perceived badness of death is that death is never undergone or experienced by someone.[21] If someone is killed painlessly in her sleep, then she may never be in a state to which she objects.

Still, Kantian constructivists can recognize several reasons why killing an individual involves violating a strict duty to that individual. I shall discuss three such reasons, and specify how they relate to killing nonhuman animals. The first two reasons refer to ways in which killing an individual amounts to harming that individual. The question of the harm of death is discussed in more detail in the first part of this volume. I shall here briefly summarize and defend two positions in these debates, and explain how such harm is morally significant in Kantian constructivism. The third reason does not refer to the harm of death, but instead builds on the Kantian constructivist views of how we should treat individuals with moral status.

10.4.2. Life as a Condition for the Successful Pursuit of Purposes

After one is killed, there is no one left to object to the condition of being dead. However, it may be argued, within Kantian constructivism, that killing is wrong because the victim, while still alive, does not agree to being killed. It may be pointed out that nonhuman animals cannot meaningfully agree or object to being killed. They can react extremely aversively when an animal or human tries to kill them, but since they have no concept of their continued existence over time, this may not be considered a relevant objection (see e.g. Singer 1999).

However, even if animals cannot have a desire to live, there is another reason why death harms them that has to do with their desires (or value dispositions). Note that biological life is generally not desired for its own sake (imagine spending the rest of your life in a coma). It is, however, absolutely necessary for the fulfillment of various purposes we have, such as finishing writing this paper or spending time with our families. This is why no rational agent, regardless of what purposes they have, can endorse a rule of killing those who have purposes they want to fulfill. Because life is absolutely necessary for successful agency, the duty against killing is quite strict (Gewirth 1978; Herman 1993).

But do animals have the relevant kind of purposes for the future? It has been argued that animals cannot be harmed by death because they do not have "categorical desires" (Cigman 1981; Belshaw, this volume). While "conditional desires" concern what we want on the assumption that we are going to be alive, "categorical desires" resolve the question of whether someone is going to be alive (Williams 1993). As an example, Julie only wants a painkiller on the condition that she is going to be alive. If Julie is dead, there is no more need for the painkiller. If Julie wants not to be in pain but also to finish writing her chapter, however, death will thwart the latter desire. According to Ruth Cigman (1981), animals cannot be harmed by death, because they cannot have such categorical desires. She claims that to have such desires, animals would have to understand that death closes off the possibility of a future for them.

I oppose the view that beings who do not have such an abstract understanding of death cannot be harmed by death (Kaldewaij 2008). It does not seem very plausible to assume that someone can only be harmed by death if he has some long-term plan that he himself understands as giving him a reason to go on living. Some people might happily live their everyday lives "in the moment," and yet still seem to be seriously harmed by death. It does appear that death is not bad if it merely frustrates desires to no longer be in some negative state, like hunger or pain. But there seems to be no reason at all for supposing that animals *merely* have such desires, and lots of reasons for supposing animals of various species positively enjoy eating food, engaging in play, socializing, and so on. They, then, require life to be able to fulfill their actual or dispositional purposes. (See Bradley, this volume, for a more detailed criticism of this dismissal of the harm of death for animals).

10.4.3. Deprivation of Future Purpose-Fulfillment

The previous view I discussed regards death as problematic because it frustrates the purposes that an agent has prior to death. But if we think about what we stand to lose in death, it also seems that death is bad because we lose parts of our future that we do not now positively anticipate. This is a "deprivation account" of the harm of death, of which Thomas Nagel (1978) is an important proponent, and which is also defended by Bradley, this volume. An example of a harm of deprivation is when someone unbeknownst to you defames your character and deprives you of a job that otherwise would have been offered to you, and that turned out to be your dream job. The deprivation view is compatible with the view that value is projected by individuals, since the idea is that beings are deprived of something they *would have* valued, in future. In the case of death, the deprivation is quite extreme, taking away your possibilities of engaging in any activity you value ever again.[22]

Someone can only be deprived of future goods if she has a relevant kind of identity over time. It must be that *this* animal is deprived of future goods *it* would have valued. As we are interested here in the animal not just as a physical entity, but also as a being who has desires and experiences, it makes sense to include a psychological component in the relevant kind of identity over time, such as psychological connectedness and continuity (Parfit 1984). Peter Singer suggests that some animals may lack the relevant type of identity because their conscious states are not internally linked over time (Singer 1999). However, note that we are here interested in animals that do not merely behave in automatic ways (stimulus-response), but that are motivated to act for a purpose. Conceiving conscious animals as having essentially unrelated experiences, such that the conscious animal would, in a way, be replaced by a successor every instant, does then not seem to be very plausible. Fish, for example (which Singer [1999] previously suspected of having no conscious states that are linked over time), are capable of learning on the basis of experience; they can learn spatial structures, for example (Gerlai 2012; Laland 2003). Persisting beliefs, desires and intentions, memory and anticipation, enduring character traits over time, etc.

are all signs of psychological continuity, and seem to be found to some extent in various kinds of nonhuman animals.

It may be, of course, that consciousness and psychological continuity and connectedness over time come in degrees. It is plausible to assume that consciousness, like all other characteristics of living beings, gradually emerged in the process of evolution. On the other hand, it is unjustified to regard nonhuman animals as overall "less conscious" than human beings because they are not aware of what *we* are aware of. After all, they naturally live in different habitats, in which different kinds of information are relevant (think of cats being acutely aware, by smell and sound, of everything that is scurrying around in the garden). However, it may make sense to think of some animals as being on the borderline of consciousness (say, invertebrates or amphibians). With regard to psychological continuity and connectedness, Jeff McMahan (2002) has argued that there may be degrees in the harm of death, based on degrees of identity over time (see also chapter 4 in this book). This might be a reason to consider some animals, especially the less cognitively complex ones, to be less deprived by death than adult humans. That would not mean that there is not a strict prima facie duty not to kill such animals, but it may be relevant when there is a case of conflict of duties (see below, section 4.5).

10.4.4. Killing as the Ultimate Denial of Moral Status

So far, I have focused on the question of how death can be seen to *harm* nonhuman animals. I would like to suggest another reason why killing an animal (or human being) is morally wrong from the viewpoint of Kantian constructivism. This is because it involves a kind of fundamental *disrespect* of another being that is supposed to be recognized as having equal moral standing in the process of determining what reasons we have for (inter) action. For a Kantian, as I shall discuss in more detail below, morality is not a matter of how much well-being or suffering there is in the world. Instead, it is a matter of how a moral agent should relate to other beings who have moral status. In killing a being, apart from the harm caused by death, we effectively *annihilate* the moral subject, and with that his values or purposes. If there is no justification for the act of killing, this is a particularly grievous form of disrespect, since we do not just disregard *some* of the victims' values, but we also deny that their values are to be taken into account *at all*. We make the other into something to use as we please, and to discard as we please. Killing, in this sense, is the ultimate denial of the moral status of the other.

10.4.5. Practical Implications

There are, then, various reasons, from the perspective of Kantian constructivism, why we should accept a strict duty not to kill nonhuman animals. We need to respect animals as purposive agents, and to not take their lives, because life is a necessary condition for the fulfillment of current and future purposes. What are the practical implications for how we may treat animals?

As mentioned above, Kantian constructivists regard the right thing as a matter of what everyone can agree with (or as guaranteeing the *conditions* for successful agency or interaction, such as life, basic health, etc.). They recognize no duty to increase well-being as such, as utilitarians do (see chapters 7 and 8), as value is projected by individuals and therefore is not agent-neutral and additive. Indeed, Korsgaard (1996c) suggests that utilitarianism presupposes *realism* about value: regarding well-being as not just valuable for its own sake *for someone*, but as having its value *in itself*. Since, in Kantian constructivism, every single individual has to be able to agree, everyone basically gets a veto against being killed. Moreover, because life is absolutely necessary to fulfill one's purposes, any agent has to accept *a strict* prima facie duty not to kill. It follows from this that killing conscious animals that pursue purposes cannot be justified on the basis that it produces more well-being than an alternative action (e.g. because the total amount of pleasure caused by eating meat outweighs the suffering and deprivations of the animals used in the process).

Kantian constructivism need not necessarily lead to an absolute ban on killing, however. Imagine that negative duties conflict; imaging, for instance, that no matter what we do, someone will be killed. When we cannot avoid violating some duty, we must try not to violate the *strictest* duty. We could, for instance, consider how necessary something is for the successful agency of the person involved (Gewirth 1978). For example, life is more essential to successful agency than not having a headache (and it does not matter just how many people will be getting the headache). When no matter what we do someone will die, we might look at the harm of death involved. The issue is not whether Sally's or Harry's death decreases the total amount of well-being more, but whether death is a greater harm *to Sally* than it is *to Harry*, making an interpersonal comparison. The harm of death, as discussed above, can vary with the degree of consciousness and the strength of identity over time, as well as with the amount of future goods that someone is deprived of. In cases where we can save only one of two individuals, this may be reason to prioritize human children over fetuses, or horses over frogs.[23]

What if killing an animal in some way is the only way to prevent a human being from being even more grievously harmed? Whether or not it would then be justified to kill the animal depends on whether strict positive duties to others are accepted. While negative duties are duties not to do something (e.g. a duty not to harm others), positive duties are duties to do something (e.g. to help others). Korsgaard (2011a) regards morality in terms of how we relate to those with whom we interact, and does not seem to recognize strict duties to help every agent that requires help. However, some other defendants of arguments from agency argue that we have strict duties, corresponding with rights, to help agents attain the necessary conditions for successful agency (freedom and well-being), if the agent cannot acquire these on her own (Gewirth 1978). If such positive duties conflict with negative duties, the strictest duty takes priority, as in when we shove someone aside to save someone's life.

What about when the harms at stake are comparable, such as lethal animal experiments designed to save human lives? First of all, it may be thought that we have *more* responsibility for what we immediately do than what we let happen, so that it is worse to kill than to let die (Gewirth 1982). Second, note that such experiments do not seem to be a straightforward case of a collision of negative and positive duties. It is not the case that *either* we kill these beings *or* the ones we are trying to save will die (as may be the case, for example, when a child soldier attacks others). We have singled certain individuals out to be used as means to save others. This seems to be ruled out by the idea of acting only in ways that others can agree with. However, if those used are thought to be so little conscious or integrated over time that they lose far less in death than those who will be saved, it may perhaps be argued that the duty to save lives here overrides the duty not to kill.

Kantians, then, recognize strict duties not to kill, unless this is the only way to avoid violating an overriding duty. The question may be posed as to whether it is *possible* for us to live without causing death to others, given facts about human nature and the world. The answer that a Kantian constructivist would give is that it may not be, but at least do what you can.

10.5. CONCLUSION

In this paper, I have argued that if Kantian arguments from agency or interaction justify substantive duties to others (including rational agents), they also justify duties to certain nonhuman animals. Only rational agents can formally determine the validity of their reasons on the basis of the constitutive requirements of (inter)action. However, they only have sufficient, substantive reasons for action if they take unmotivated desires, or subjective values more generally, as a prima facie valid basis for (inter)action. Some nonhuman animals can plausibly be thought to act on the basis of nonlinguistic desires for objects or activities. We must therefore also respect their purposes, and life is the necessary precondition for current and future purpose-fulfillment. This leads to a strict duty not to kill nonhuman animals, unless this is the only way to avoid violating overriding duties.

ACKNOWLEDGMENT

This article is part of the research program "Human Dignity as the Foundation of Human Rights?," which is financed by the Netherlands Organisation for Scientific Research (NWO). My project partly deals with the question whether only *human* beings have dignity and rights.

NOTES

1. Kant claimed that we have a duty to *ourselves* not to be cruel to animals, as this weakens our capacity for sympathy, which helps us do our duty to other rational beings (2000, 6:442–3). See Denis 2000 and Kain 2010 for interpretations of this view. For the view that Kant regards the capacity for moral autonomy (or "pure practical reason") as that which gives rational beings a special status, see, for example, Timmermann 2006. Richard Dean, controversially, interprets Kant as claiming that a morally good will is required for the status of an "end in itself" (2006). For the view that broader capacities of practical reason are required for moral status, see, for example, Korsgaard 1996a, Wood 1999.
2. Indirect duties can, of course, be strict duties owed to the *self*. Lara Denis argues that Kantian indirect duties to animals imply more extensive restrictions on our dealings with animals than is commonly thought. For example, she notes that slaughtering animals may erode our moral sentiments, and should therefore only be done to "preserve or substantially further" rational nature, such as when eating meat is a necessary or especially good means to secure nutrition for rational beings (Denis 2000, 412–14). Even if duties regarding animals are stricter than is commonly thought, however, it seems to make sense to give (negative or positive) direct duties to rational beings priority when they are in conflict with indirect duties regarding animals. After all, respect for rational nature is the ultimate point of the duties. In contrast, I argue in this paper that, in Kantian constructivism, certain nonhuman animals should be regarded as possessing the basis of a *valid claim* on rational beings, which means that we owe them certain treatment for their own sake.
3. Whether this agreement is taken as actual or hypothetical, and whether and which specific conditions are stipulated, varies with the different arguments. Also note that some authors do not strictly regard the moral principle in terms of

respect for the *reasons* or *purposes* of agents, but in terms of respecting the generic *conditions* for successful agency or agreement.

4. Sharon Street (2008) and Christine Korsgaard (2009) both make the claim that certain principles are constitutive of agency. They claim that one is not truly acting or interacting unless one acts in accordance with formal principles that, so to speak, make up the "rules of the game" of these activities. There is also a notable tradition of referring to similar arguments as "transcendental arguments" (Illies 2003, 30–32; see also e.g. Korsgaard 1996d, 123–24). (Such arguments are sometimes also called "reflexive" or "retorsive.") Such arguments posit that we are rationally required to accept some *y* because it is the necessary precondition of the possibility of some *x*. This *x*, in turn, is either a minimal, neutral premise we all accept (e.g. that there are reasons for action), or the judgment of a skeptic, who is supposed to commit himself to the truth of the judgment he denies, because in questioning it, he performatively commits himself to it (Illies 2003, 31–2).

5. Sharon Street (2008, 2012) offers a Humean global constructivist theory.

6. An author who pursues a similar strategy, without actually identifying himself as a Kantian, is Alan Gewirth (1978). Gewirth argues that any agent necessarily has to claim rights to the necessary conditions of successful agency (freedom and well-being), on the sufficient ground that he is an agent, and therefore has to attribute similar rights to other agents.

7. I have explained Kantian constructivists as regarding reasons in terms of what everyone involved can accept. It may be noted that pure arguments from agency regard reasons as a matter of what *I*, as an agent, have to accept, while only arguments from interaction understand the right thing in terms of what *we* can *all* accept. However, in arguments from agency, the generation of norms idiosyncratic to the specific agent is taken to be avoided by asking not what I can accept as the specific individual with the specific ends I have, but what I can accept *as an agent, regardless* of what specific ends I have. It is thought to follow from these arguments that all agents need to respect anyone's purposes, or the generic conditions that are required for successfully pursuing them (e.g. freedom and well-being). So I take the end result of the arguments (if they are valid) to be similar: the right thing to do is what everyone can agree with.

8. Gewirth (see note 7) takes the latter course: we must necessarily will the conditions of successful agency, which includes, for example, noncoercion (so that your actions are voluntary), and well-being (to be able to fulfill your purposes). Barbara Herman (1993, 125–26) interprets Kant along similar lines.

9. Korsgaard's theory also includes an argument from interaction. Note, however, that Korsgaard holds that reasons are grounded in the authority the human mind has over itself, not in interaction (2007, 11). Even considered as solitary agents (or when we "interact with ourselves"), we have to be able to universalize our maxims, and so already have to accept duties regarding others (2009, §9.7). Korsgaard does hold that we need an argument from interaction to show that we owe duties *to* others, not to the self (2007, 9).

10. I mention Darwall's theory here as an interactive form of global Kantian constructivism, but Darwall does not hold a constructivist account of all reasons (Darwall 2007, 60), and says that he wants to keep the answers to metaethical questions open, even though he would happily embrace constructivism (Darwall 2007, 64).

11. For example, Apel (1980) argues that nobody who acts or argues in a meaningful way can avoid entering into a rational debate with all other rational beings, and taking all contributions seriously. Habermas (1990, 96–102) disagrees and thinks

that rejecting moral norms is rationally possible, but that individuals who really tried to act in a purely strategic, and not a cooperative way, would become isolated and schizophrenic. Darwall seems to have a position somewhere in between. He argues that we cannot psychologically avoid feelings of resentment. Darwall does not argue that we have a "first-personal reason" for accepting moral duties, but he does argue that, once we accept "second personal" reasons, we have no reason to consider them any less valid than "first personal" ones (2006, 277–300). He also says that we unavoidably take up a second-person perspective when we respond to someone's making a claim on us as a "summons" (2006, 59–60).

12. I have worked these arguments out in detail in my dissertation, "The Animal in Morality: Justifying Duties to Animals in Kantian Moral Philosophy" (Utrecht University, 2013).

13. Korsgaard works out her arguments concerning the moral status of animals in, amongst others, the following published works: Korsgaard 1996d, §4.3; Korsgaard 2005; Korsgaard 2011a. At the time of writing this paper (October 2014), she also has several works in progress on this topic, and on the natural origin of the good. These documents (including Korsgaard, C. forthcoming, 2011b and 2011c and 2011c) can be downloaded from her website, at http://www.people.fas.harvard. edu/~korsgaar/Essays.htm.

14. I am not sure that Korsgaard would agree with my claim that the actions of rational beings fundamentally have a similar basis as the actions of certain nonhuman animals. She argues that "while it is our rational nature that enables us to value ourselves and others as ends in ourselves, *what* we value, what we declare to be an end in itself, includes our animal nature as well as our rational and human nature" (forthcoming). So she understands the *content* of human action to be similar (at least sometimes), but that does not mean the *basis* of human action is the same. Korsgaard does sometimes say that we choose things "because they are important to us" or "because we want to do it" (e.g. Korsgaard 1996d, 122; Korsgaard 2011b, 24), which is in line with my view. However, a central idea of Korsgaard is that rational beings are free in a way that animals are not; they are able to choose their principles for action and constitute their own practical identity (the central idea of Korsgaard 2009). As I shall argue below, while I agree that rational beings are free to reject or endorse reasons on the basis of the constitutive requirements of action and interaction, I think that the reasons that are the output of these procedures are still ultimately based in desires or values agents have not chosen. This means that, in a way, human beings do not constitute their own identities, but discover what they really entail. It may be that the view that human beings have a capacity for free choice over and above that of animals is a crucial part of Korsgaard's argument for moral duties to others. If so, I think this makes the argument problematic. I cannot work this out further here, and assume the general validity of these Kantian constructivist arguments (specifically, the idea that action or interaction implies sharing reasons with everyone, or all concerned).

15. Korsgaard explains the agency of some animals in terms of acting on the basis of evaluatively or motivationally loaded representations (Korsgaard 2011b, 7–16; Korsgaard 2009, 90–108). In her view, nonhuman animals do not just act on such incentives, but also act on a kind of principles, though unlike the principles of rational animals, these are not freely chosen.

16. Instead of calling purposive agency without the capacity for formal reasoning "prerational," we might also understand it as being rational in another way (e.g. in line with someone's beliefs and desires). However, to avoid confusion, I shall

only use the term "rational" for "being endorsed or not rejected on the basis of constitutive principles or procedures."

17. Some may believe that this argument is "un-Kantian" in regarding the moral procedure as obligatory for an agent *regardless* of the purposes she pursues, rather than *independently* of any purpose she may have. Kant, after all, regarded the categorical imperative as *unconditional*. It may be thought that the moral test has to be a pure requirement of reason. Kant's test of what we can "think" as a universal law, understood as avoiding *theoretical* contradiction, may be thought to be a good example. (If one understands it as a *practical* contradiction, the contradiction lies in that the agent's maxim ceases to be efficacious in a world in which it is universalized: no one will believe your lie if everyone lies [Korsgaard 1996b, 78]. The duty to tell the truth is then still conditional on wanting to achieve your subjective end, or whatever you hope to achieve by lying.) Note that, according to the theoretical contradiction test, a reason such as "I will coerce or manipulate others for the sake of my self-interest" is wrong not because it undermines someone's possibility of free agency, but because it is logically inconceivable as a law for all moral agents (as coercion or manipulation becomes impossible if everyone does it). It is then wrong in a way that does not appear to be interestingly different from making a mistake in theoretical reasoning. But most importantly, even if this test does exclude some maxims in a relevant way, it cannot give us reasons for action, which is the premise of arguments from agency. It tells us not to lie, but it does not offer a basis for endorsing a policy of telling the truth. If I have reason to communicate, then I must speak the truth. But the reason to endorse communicating in the first place must come from elsewhere than pure reason alone.

18. While I cannot go into the specific different variants of the arguments in this text, I do at this point briefly want to explain why my argument also applies to O'Neill's form of Kantian constructivism. It may be noted that O'Neill emphasizes that the procedure she advocates does not ask whether agents *would* (in ideal circumstances) or *do* accept certain principles, but whether they *can* accept them (e.g. O'Neill 1989, 217). She says principles must be "followable" in both thought and action. This may seem to be a minimal formal rational requirement, but I would argue that it involves more. For example, it includes that certain principles not strike agents as arbitrary, from their points of view (O'Neill 2002, 359). But this seems to suggest that agents must at least be *able* to accept the principles *from their own (subjective) normative or evaluative starting points* (and their metaphysical beliefs, etc.), even if they do not actually accept these principles. Otherwise, it is difficult to see from where the "arbitrariness" of claims would be decided.

19. Korsgaard agrees with Darwall that we need to be under "common laws" to be able to owe something to each other: if I autonomously will a law that says that you should be treated well, then this does not establish that I owe this treatment *to* you (2011c, 28). She argues that while animals are not "active citizens" (legislators), they are "passive citizens," like women were in times before women's suffrage (2011c, 31). Such beings fall under the protections of laws others legislate, and can claim it as a right from those others to treat them in certain ways, because they have a property in virtue of which we ourselves would claim it as a right from another (2011c, 32). Why can animals claim certain treatment of us as a right, and therefore be not just objects under the law, but "citizens" of a kind? In the main text, I try to explain this in my own terms.

20. Here is another point where my argument differs from Korsgaard's. I think we should not morally respect the goods of "active citizens" (see note 20) because

they have *legislated* them to be good (Korsgaard 2011d, 32–33; see note 13 above), but simply because they value them: on the same basis why we should respect the goods of "passive citizens." We should respect the rational capacities of others because our own rational capacities are fallible and we may make mistakes when we reason formally, but that is not the core of moral duties to others. Respecting someone's capacities of reasoning is morally relevant, however, because denying that someone has the capacity to reason may be a way to make sure their subjective values are not taken seriously.

21. Epicurus already pointed out that, when we are alive, we are not dead, and when we are dead, we do not exist (Rosenbaum 1993, 121).

22. Of course, it may well be that animal lives are not worth much to them because these lives are very miserable, but that cannot be used as a reason for killing them if they are in this state because we violated previous duties to them (e.g. if we failed to treat the animals in our care well).

23. To explain: embryos are not conscious yet, so they are not psychologically continuous or connected to the person who will grow out of them, and horses may be more conscious and more diachronically unified than frogs.

Killing Animals and the Politicization of Normative Ethics

PART III
Killing Animals and the Politicization
of Normative Ethics

11

Life, Liberty, and the Pursuit of Happiness? Specifying the Rights of Animals

ALASDAIR COCHRANE

Rights are usually taken to have a quite distinctive and powerful role in ethical thinking. First of all, rights are generally taken to embody constraints on the pursuit of social utility, and they have thus been described as "firewalls," "side-constraints," and "trumps" on purely utilitarian reasoning (Habermas 1996, 258–59; Nozick 1974, 28–29; Dworkin 1984). As such, theories of animal rights obviously rival utilitarian accounts of animal ethics, such as that defended by Peter Singer. But they also do much more than that. For rights also possess an intrinsically *political* character; that is to say, by identifying rights, we do not merely outline what is morally good, desirable, or recommended, but instead specify a set of entitlements and correlative duties that can usually be demanded and coercively enforced by the state (Nozick 1981; Steiner 2005). In this way, animal rights offer a huge challenge to conventional political thinking and practice, for they not only demand certain basic entitlements for animals that cannot easily be overridden, but they also stipulate that those rights and duties cannot just be left to the good will of individuals, but are a central concern of *justice*. As such, the consequences of recognizing and upholding animal rights are potentially immensely radical. Indeed, given the amount of animal killing that all human societies presently undertake, the consequences of recognizing an animal right to life would be profound.

But can nonhuman animals possess rights? And if so, just which ones do they possess? The former question dominated early debates in animal ethics, and since this topic is already much discussed in the literature, it is not this paper's main focus (see Frey 1977; McCloskey 1979; Regan 1976; Feinberg 1974). Instead, the paper will concentrate on the much more neglected latter issue: How we can specify the rights of different animals in a plausible and determinate way? This paper identifies and evaluates three existing means by which to specify the rights of animals: by their inherent value, by their important interests, and by their relational position. It argues that while each of these accounts has something to offer, including to the question of animals' right to life, none on its own offers a convincing theory for delineating animal rights. Instead, a two-tiered interest-based approach is briefly outlined and defended, and its implications for the right to life are explored. Under this approach, the "abstract" rights of animals are specified by judging which of their basic interests are sufficient to ground duties in others. These abstract rights are then translated into the more specific "concrete" rights of a particular animal in a specific situation, through a careful analysis of all the relevant interests and values at stake in that context.

11.1. RIGHTS AND INHERENT VALUE

The first and perhaps most famous means by which to specify the rights of animals comes from recognizing them as beings with "inherent value." This view claims that once we recognize that certain animals have this inherent value, we necessarily also recognize that they possess specific rights. This position is adhered to by probably the two most well-known animal rights theorists, Tom Regan and Gary Francione, and both share the understanding of inherent value proposed by Regan in his book *The Case for Animal Rights* (2004). In that book, Regan argues that some animals are "subjects-of-a-life." That is to say, some animals have beliefs, desires, a sense of themselves over time, interests in their own fate, and so on (243). Regan claims that creatures who possess these mental attributes and conscious awareness have an "inherent value", or a value all of their own, which cannot be reduced to how valuable or useful they are to others (243).

Crucially, Regan also claims that inherent value establishes one fundamental and basic animal right: the right to "respectful treatment" (276). This right amounts to an entitlement to always be treated as the kind of being who possesses inherent value, and thus never to be harmed simply as a means to securing the best overall consequences. Regan argues that using individuals "merely as a means" to aggregate utility treats them solely as "receptacles" of value, as opposed to recognizing that they have value in and of themselves. Furthermore, Regan also claims that a number of other animal rights are entailed by this right to respect, including a general right not to be harmed: "We fail to treat such individuals in ways that respect their value if we treat them in ways which detract from their welfare—that is, in ways that harm them" (262). Moreover, Regan is also quite clear that death is the "ultimate harm" that can befall animals, on the basis that it "forecloses *all* possibilities of finding satisfaction" (100). As such, Regan also believes that his theory encompasses the right to life.

However, the notion of inherent value cannot establish all of the rights that Regan wants from it. For, in fact, neither the right not to be harmed nor the right to life are entailed by the notion of inherent value or the right to respectful treatment. After all, the point of the right to respect, you will recall, is that it protects an individual from being harmed *simply to secure the best overall consequences.* This is certainly entailed by the notion of inherent value, because when we use someone solely to secure benefits for others, we fail to treat that individual as if she has a worth of her own. But quite obviously, not all harms—including the harm of killing—are intended simply to secure the best overall consequences. In fact, many of the most significant harms we condemn—such as sadistic wanton acts of cruelty against companion animals, or the trophy hunting of members of endangered species—can quite plausibly be said to result in lower overall utility, all things considered. Moreover, it is perfectly possible to harm and even kill an individual while also recognizing that he has value of his own. One pertinent example—which Regan himself recognizes—is self-defense. It is consistent to recognize that an aggressor has value that is not reducible to his value to others, while also causing that individual serious harm, perhaps even death, through self-defensive action. As such, neither inherent value nor the right to respect, as Regan understands the terms, necessarily entail a general right not to be harmed or a right to life.

One might respond that it is simply obvious that inherent value entails the right not to be harmed and killed. After all, it might be argued that to have this kind of value is to be an equal within the moral community—which itself entails certain equal rights, such as to life and liberty (Jamieson 1993, 224). But even if the possession of inherent value does entail a certain equality—which is not self-evident—it would be extremely odd if it entailed a set of rights identical to that possessed by all other members of that community. After all, Peter Singer (1986) has persuasively argued that moral equality does not mean the same thing as equal treatment; individuals can be moral equals, but nevertheless be owed quite different things.

Children, for example, are rightly considered to have equal moral status or value to adults, and yet they justifiably possess a quite different set of rights.

So the question of whether animals have the right not to be harmed or the right not to be killed cannot be settled simply by invoking such notions as inherent value or equality. These concepts are, of course, powerful and useful, but on their own they cannot generate a comprehensive account of the rights of animals. But it is little wonder that Regan tries to incorporate harm within his account of animal rights. After all, it does seem correct that rights are fundamentally about protecting individuals from harm; that is to say, they are about protecting individuals' *interests*. In order to properly specify the rights of animals, then, the most obvious place to start is not with the notion of inherent value, but with an account of the interests of animals.

11.2. RIGHTS AND INTERESTS

Interest-based understandings of rights were of crucial concern in the early debates in animal ethics concerning whether animals possess rights (see Frey 1977; Feinberg 1974; Regan 1976). Thinkers like Joel Feinberg argued forcefully that the necessary and sufficient condition for the possession of rights is not moral agency, or the ability to claim rights for oneself; rather, it is the capacity for interests. For when an individual has interests, then those interests can be represented and claimed by a proxy on his or her behalf. Furthermore, Feinberg argued that since sentience is the necessary and sufficient condition for the possession of interests, all sentient animals possess rights.

Of course, the very fact that animals can possess rights does not on its own tell us anything about just *which* ones they might possess. For this, we need a robust account of what the interests of animals actually are, and we also need an account of how those interests can be translated into rights. For the latter, it is possible to borrow from Joseph Raz's famous formulation of rights: "'X has a right' if and only if X can have rights, and, other things being equal, an aspect of X's well-being (his interest) is a sufficient reason for holding some other person(s) to be under a duty" (1988, 166). We can use this Razian formulation, then, as a starting point for specifying the rights of animals. Not all of the interests that animals possess equate to animal rights. Rather, we can specify the rights of animals by determining which of their interests are sufficiently strong to ground duties (positive or negative) on the part of others (Cochrane 2012a).

One of the biggest problems facing this account in terms of helping us specify the rights of animals is coming up with an accurate account of what the interests of animals actually are. For example, this volume demonstrates that the debate over the seemingly simple question of whether animals have an interest in continued life is fiercely contested. However, simply because the task is demanding, that does not make it hopeless. There are similar debates about the basic interests of human beings—including over the question of continued life—but the interest-based approach nevertheless remains the most popular means by which to specify human rights (see Griffin 2008; Tasioulas 2002; Caney 2007). The answer to this problem does not lie with abandoning the interest-based approach, but with careful argumentation. First, it is necessary to have a proper understanding of what precisely the concept of "interest" entails. And second, to determine the content and strength of the interests of animals, it is necessary to undertake a rigorous examination of what capacities, traits, and behaviors different animals possess.

While there is not space in this paper to conduct either of these tasks in full, it will nonetheless be useful to sketch out, in a very preliminary way, how such argumentation might take place. First of all, then, there is the question of what an interest is. Following Feinberg (1984), and most ordinary understandings of the term, I take interests to be components of *well-being*. To have an interest is to have some kind of *stake* in a good, and to gain or lose depending on the condition of that good. Importantly, then, for an individual to have an interest in something does not mean that she actually has to be interested in that thing, or even to desire that thing; it simply means that the presence or absence of that thing advantages or disadvantages her in a way that matters to her (Cochrane 2012a). This means that nonsentient entities like plants possess no interests. And it also means that sentient individuals can have interests in things that they have no concept or understanding of—such as a baby's interest in breathing clean air.

Turning to the question of the content and strength of animal interests, at the most general level we can say that the behavioral, physiological, and neurological evidence all strongly suggests that there exists a set of nonhuman animals who are sentient beings, meaning they are aware of themselves and their surroundings, and they can feel pleasure and pain. In other words, there exists certain animals with the capacity for well-being, and thus who possess interests. Furthermore, we can also say with some confidence that as sentient beings, some of the strongest interests these animals possess include avoiding pain and pursuing pleasurable experiences. Given that we also know that these animals are not "stuck in the moment," but possess some future-oriented desires, we can also say that this ordinarily gives them an interest in satisfying future-oriented goods—and thus in continued life itself.

Having established these interests in avoiding pain and in continued life—as well as other interests that there is not the space to explore here—we then need to ask whether they can establish corresponding *rights* not to be made to suffer and not to be killed. Are these interests "sufficient" to ground corresponding duties? At an abstract level, it certainly seems like they can. After all, the interest in not suffering is strong and pressing. It would be astonishing if it did not impose at least a prima facie duty on us not to respect it. We obviously have a duty, for example, not to cause serious suffering to animals simply for our amusement. And that duty is established because such actions set back an important interest of the animals themselves. Given that rights, according to the interest-based approach, are those interests that are sufficient to impose duties, then this means that animals do have a prima facie right not to be made to suffer.

The case of continued life might be thought to be different—and to depend on the animal in question. After all, it is reasonable to assume that a good many sentient animals possess only a very weak interest in continued life compared to more cognitively complex creatures. For example, while birds do have future-oriented desires, it is clear that what McMahan calls their "psychological continuity" over time is weaker than that of certain other animals, like human beings (2002, 233). Because birds have a weaker sense of themselves over time, less ability to conceive of themselves into the future, and fewer future-oriented goals than most human beings, for example, it is only reasonable to say that their strength of interest in continued life is correspondingly weaker. But still, that bird interest must still be sufficient to ground a prima facie right to life. For example, I regularly come across birds when I walk home from work, and it does seem that I have an obligation not to grab and beat one to death just because I had a bad day in the office. Moreover, my duty to refrain from such action derives from the bird's own interest in continued life. Once again, then, this shows that birds do have at least a prima facie right to life. That right to life may be weaker than that possessed by more cognitively complex creatures, and so more easily overridden in certain circumstances, but that does not mean that the right itself should not be recognized.

At this point, however, some will argue that the kind of analysis presented above cannot provide a comprehensive account of the rights of animals. It might be argued that by focusing solely on definitional questions and issues surrounding animals' particular capacities, interest theories are blind to the importance of *social relations* in specifying rights. This critique is put forth most forcefully by Elizabeth Anderson (2004), who argues that some rights—whether they be human or animal—depend not merely on the capacities of the individual in question, but also on whether they are a *member* of society, and on the possibility of *peaceable relations* between the right-bearers and the individuals with the corresponding duties. Taking the membership point first, Anderson asks us to imagine a pod of dolphins in the ocean who would starve if we did not feed them, because their usual food source has collapsed. She argues that although those dolphins certainly have a strong interest in being fed, they cannot properly be considered to have a *right* to be fed by us, on the basis that rights to positive assistance are tied to social membership. Anderson is not arguing that dolphins lack any rights at all, nor is she arguing that no animal ever has rights to positive assistance. Wild dolphins possess the right not to be harmed under her account, and domesticated and captive animals merit positive provision on the basis of their membership of human society. Her point is that to properly determine the rights of animals, we need not just an account of the interests of animals, but also of their social membership.

Indeed, for a full account of animal rights, Anderson also believes that we need to know whether there is a possibility of peaceable relations between us, moral agents, and the animals in question. She gives the example of certain species of rats and mice who have found their ecological niche inside our homes. It is evident that such animals have clear interests in not being interfered with, ejected from our homes and in not being killed. And yet, Anderson argues that these types of animal have no such rights. This is because "vermin, pests, and parasites cannot adjust their behavior so as to accommodate human interests. With them there is no possibility of communication, much less compromise. We are in a permanent state of war with them, without possibility of negotiating for peace. To one-sidedly accommodate their interests . . . would amount to surrender" (288). Once again, Anderson's point is not that such animals have no rights—torturing rats for our amusement would, she argues, constitute a violation of their rights. Rather, her point is that an account of the most important of these animals' interests is insufficient for specifying their rights.

One problem with Anderson's approach, however, is that it remains underspecified. For example, she regards the distinction between animals that reside within our societies and those that do not as clear cut. But while it is relatively straightforward to identify wild dolphins as outsiders and companion dogs as full members of our societies, there are a huge number of animals who occupy more problematic intermediate positions. For example, the nondomesticated animals who live among us and often depend upon us, such as birds, foxes, hedgehogs, rabbits, and so on, are partly members of our societies, and yet it is unclear what kinds of rights they would be afforded in Anderson's account. She might argue that they are "vermin" on the basis that they have found their ecological niches within and around our dwellings and are incapable of adjusting their behavior to accommodate human interests. But this would be highly counterintuitive: our obligations to garden birds surely extend beyond not torturing them for our own amusement.

In sum, while Anderson is correct to point out that in order to specify the rights of animals we need more than simply an account of the interests that they have, her own means of delineating those rights is problematic and incomplete.

11.3. RIGHTS AND RELATIONAL POSITION

A much fuller relational account of animal rights has been proposed by Sue Donaldson and Will Kymlicka in their book *Zoopolis* (2011). Donaldson and Kymlicka argue that an interest-based approach focused solely on the capacities of animals can ground a very thin set of animal rights, including the right to life. They call these the "universal basic entitlements" of animals. They also agree with Anderson that a fuller account of animal rights demands a focus on the different relational positions of animals. However, not only do Donaldson and Kymlicka provide a far more systematic account of the particular rights that specific animals possess, but they are also alive to the difficulties in delineating animals' relational position.

To explain, Donaldson and Kymlicka (2011) argue that domesticated animals should be viewed as members of our shared society and should enjoy membership rights as such. In particular, these rights can be thought of as the rights of *citizenship*, which include rights to political concern, to political agency, and to residency. Wild animals who live apart from humans, on the other hand, should be recognized as being competent in running their own affairs without intervention. As such, these animals should be granted rights to *sovereignty*, which include rights to their own territory and autonomy on that territory. Finally, they argue that those nondomesticated "liminal" animals who live among us should be granted rights of *denizenship*, which include rights of residency, but which fall short of the full rights of citizenship. In this way, Donaldson and Kymlicka claim to provide a full and robust specification of animal rights: all sentient animals possess certain basic universal entitlements grounded in their basic interests, including the right to life, but they also possess differentiated rights based on their particular relational position.

But while Donaldson and Kymlicka's theory should be commended for its attempt to provide such a comprehensive account of the rights of animals, it falls short in providing a convincing account of how rights ought to be specified. In particular, it is not evident that we can plausibly carve animals up into three relational groups—domesticated, wild, and liminal—each of which possesses a distinctive set of group-differentiated rights (Cochrane 2013). The way in which they treat the issue of the right to life is illustrative of this concern. You will recall that they claim that all sentient animals have a basic universal entitlement to life. However, because of their different relational positions, they argue that what this right amounts to for wild, liminal, and domesticated animals is very different. They claim, for example, that for domesticated animals this entails a duty on our part not just to protect animals from humans, but also to protect them from other predator animals. However, they are quite clear that when it comes to liminal and wild animals, we have no such duty to offer protection. This is because protecting liminal animals would come at drastic cost to their liberty, which there is good reason to believe they would not desire. Furthermore, while domesticated animals ought to be considered as members of *our* communities and thus entitled to the forms of basic absolute individual protection that membership affords, wild animals ought to be considered as members of their *own* communities, which can only flourish if natural food cycles and predation are allowed to continue.

However, while Donaldson and Kymlicka's desire to refrain from protecting liminal and wild animals from predation has a good deal of intuitive appeal, matters are more complex than they admit. After all, we do in fact already recognize that some wild and liminal animals have rights not to be killed by predators. The most obvious examples concern the protection we afford to wild animals from *human* predation through limitations and bans on hunting, such as the international moratorium on whaling. But it is also relatively common to protect wild animals who belong to rare species from "overabundant" predators, as in efforts made in parts of North America to protect Blanding's turtles from raccoons. And, of course, many of us also believe that the liminal birds, rodents, rabbits, and other animals that domesticated cats and dogs prey upon are worthy of protection and should not be left at the mercy of our companion animals (Cochrane 2013a).

The point here is that while relational position may be relevant in determining certain rights, it is not all that is relevant. Other factors, such as the capacities and abundance of the animals in question, as well as the relative burden of the duties entailed by the putative rights, are also crucial. As such, it is overly simplistic to divide animals into three relational categories, each with its own distinctive rights. The right to life and the right to be protected from predation can both extend beyond domesticated animals. This reveals that factors other than relational position are necessary for a full and compelling specification of the rights of animals.

11.4. A TWO-TIERED INTEREST-BASED APPROACH

So far this paper has reviewed three existing positions for specifying the rights of animals and has found them all incomplete. In this final section I want to propose and briefly sketch an alternative means by which to delineate the rights of animals; a means which avoids the pitfalls of alternative approaches, but which also aims to draw on their strengths. The approach that I propose adapts and borrows from a "two-tiered" theory of human rights that has been sketched by John Arras and Elizabeth Fenton (Arras and Fenton 2009, Cochrane 2012b). As its name suggest, this approach specifies rights at two levels.

The first level specifies a set of "abstract" or "prima facie" rights. These rights mirror the types of rights written into declarations and bills, and are incredibly useful in pointing to the general conditions of a minimally decent life. Moreover, they are politically powerful in providing social movements with a resource with which to agitate for change. These rights are specified via the basic methodology of the interest-based approach outlined above. First, the most important interests of animals must be identified, which requires an understanding of what the concept of interests entails, as well as a detailed understanding of animals' particular individual capacities, traits, and behaviors. If a basic interest of an animal can be shown to be pressing enough to impose a duty upon another, then it establishes an animal right. The rights established in this way can be regarded as "abstract," in the sense that they do not outline precisely what any particular animal is owed and by whom. And these rights are "prima facie," in the sense that they are defeasible and may in particular situations be outweighed by competing considerations (Cochrane 2013b). This "tier" of rights outlines the goods that animals require to lead minimally decent lives, but on its own it is unlikely to tell us what any particular animal ought to get in any particular situation. For a full specification of what these rights amount to, we also need to identify a second tier of "concrete" rights.

Concrete rights emerge out of the abstract tier outlined above, but they provide detail and specificity as to what those rights actually amount to in particular contexts. In this sense, they mirror the rights that are determined by specific court judgments, legislative processes, and other institutional mechanisms. This process involves resolving conflicts between abstract rights and other rights or values, and identifying the specific duties that are owed in particular rights claims. This is done not only by careful consideration of the important interests of the animal in question, but also of a whole range of other contextual factors. Importantly, those factors will include, but will not be limited to, the animal's relational position.

To give a sense of how rights might be specified in this account, let's return to the question of the right to life and consider the example of rodents. Rodents are particularly useful to consider, given that we have already discussed them in relation to Anderson's relational theory. How would the two-tiered interest-based account resolve issues concerning rodents' right to life?

When it comes to their abstract or prima facie right to life, the issue is relatively straightforward. The case was already made above that as sentient animals, these creatures have future-oriented desires and goods that continued life affords them to pursue and realize. As such, it can only be concluded that rodents do have an interest in continued life. Furthermore, it also only seems reasonable that very often this interest is sufficient to ground duties on us not to kill them. Once again, it would seem quite wrong for me to grab a squirrel that I pass on my way home for work and beat it to death. And once again, my duty to refrain from such action derives from the squirrel's own interest in continued life. As such, it makes sense to say that rodents at least have an abstract and prima facie right to life.

But what of rodents' concrete right to life? Of course, since concrete rights are dependent on context, and not every context can be examined here, only a sketch of an answer can be given. But let us explore one of the most difficult cases: the situation of those rats and mice who live around our dwellings and sometimes attempt to take up residence within our homes. They may have an abstract right to life, but what concrete rights do they possess? The first important point to make is that we cannot, as Anderson does, group all animals within a specific pejorative category—such as "vermin"—and ignore their vital interests. We have already established that these animals have important interests that impose duties on us. Those duties may not be grounded in all contexts, but since that interest is important and sometimes pressing enough to ground duties, we must assume that the duty holds until compelling reasons are presented for overriding it. Anderson claims that one compelling reason for overriding it is that we cannot have "peaceable relations" with such creatures. But this is not true. For in fact we can and do have peaceable relations with such rodents—we can keep them from our homes and property through proper building and sewer maintenance, and through the proper storage of food and waste. Of course, sometimes efforts to prevent rodents gaining entry to our homes will fail or simply be too late. In these contexts, then, it might seem as though these rodents lose their concrete right to life. But even this is too hasty. After all, live trapping is possible, and if the traps are checked regularly, and if the rodents are returned to familiar territory, then there is no reason why the animals cannot flourish.

And yet sometimes live trapping may not be possible. For example, releasing rodents nearby can be unsuitable when an infestation in a specific area means that damage to and invasion of our property is inevitable. Furthermore, relocating animals elsewhere can also often lead to their suffering and death, because of a failure to adapt to life in a new territory. In these situations, then, given that rodents are the sources of many human pathogens that cause serious diseases for human beings (Begon 2003), perhaps we can say that killing these animals is a matter of legitimate self-defense. If so, their concrete right to life would not be grounded in that particular context.

But still this is too quick. After all, if the problem is that there are so many of these rodents in a specific area that invasion and damage is inevitable, it is not clear that killing these animals will reduce their numbers effectively. In fact, it is quite possible to reduce their numbers through developing and administering contraceptives, as opposed to poisons (Flegenheimer 2013). Crucially, there is evidence to suggest that administering contraceptives is a far more effective means of reducing numbers over the long term in an overpopulated species compared to killing. After all, a swift cull simply results in a better environment for the remaining population to thrive and breed in (Zhang 2000, 106). It is thus reasonable to conclude that when these rodents are overabundant, we have a duty to administer contraceptives rather than kill them. As such, their concrete right to life still holds.

At this stage it will be pointed out that effective contraception for rodents remains rare, not least because companies have lacked any incentive to develop the relevant rodent contraceptives, given societies' preferences for extermination. However, there is no intrinsic problem with developing effective contraception for rodents, and when incentives have been in place—such as to protect farmlands or transport networks—modest successes have been achieved (Watts 2009). The more serious difficulties lie in administering the contraceptives to sufficient numbers of rodents so as to make a real difference to their numbers. Given the rapid reproduction rates of rodents, administering the contraceptives to enough of them to bring down their population is currently very difficult. We might then conclude that in situations where rodents have invaded our property, and where the threat they pose cannot be prevented by nonlethal defensive action, trapping, or contraception, the killing of such rodents would be permissible. In other words, in those particular contexts, rodents would not possess a concrete right to life.

Some will object that even in these contexts the rodent right to life should still stand. After all, it might seem that killing these animals would be disproportionate, given that the pathogens they carry rarely lead to the *deaths* of human beings. However, it must be remembered that death is not ordinarily equally harmful to rodents and humans. Because of their reduced cognitive complexity, the rodent interest in continued life is ordinarily much weaker than that of human beings, and so death is less of a harm to them. Furthermore, the human interest in avoiding serious disease is powerful indeed. As such, it only seems reasonable to conclude that the threat of a serious harm short of death—like the transmission of a serious disease—is a sufficient basis to override our obligation not to kill rodents when all of our other options have run out.

But even in these contexts, self-defensive killing cannot take any form. As sentient beings, rodents also have a powerful interest in not being made to suffer. And this interest must impose duties upon us in situations of justified self-defense. After all, there are a range of means by which we might kill animals that have invaded our properties. The most common current method of killing rodent pests is through the use of anticoagulant poisons, which are known to cause serious suffering (Mason and Littin 2003). Given that more humane alternatives are available and effective, invasive rats and mice have a concrete right that we use them.

This discussion has by no means provided a comprehensive account of the rights of rodents—or even of the rights of rats and mice who seek to invade our homes. Much more can be said about the other abstract rights of these animals, and different contexts will lead to the attribution of different concrete rights. The aim of the discussion was simply to illustrate how the two-tiered method specifies the rights of animals. The first tier identifies a set of abstract rights that are derived from the most important interests of animals. This focuses our attention on the conditions required for a minimally decent life for animals, and establishes general duties that can only be overridden for compelling reasons. The second tier specifies more precisely what any individual's abstract rights amount to in specific circumstances. This is undertaken through a careful and thorough examination of all the relevant interests and values at stake in any particular putative rights claim.

11.5. CONCLUSION

Various approaches to specifying the rights of animals have been reviewed in this paper—and all have something important to offer contemporary debates in animal ethics, and on the question of animals' right to life. The notion of inherent value is important, but on its own it cannot tell us which specific rights animals possess, including the right to life. Interest theories do better in that task by linking animal rights to their particular capacities and the important interests that flow from them. However, they ordinarily fail to acknowledge that social relations and context can impact upon the entitlements of animals. Relational theories aim to remedy that fault, but have ended up fetishizing discrete groups of animals, failing to acknowledge that different rights—including the right to life—can exist within a particular group, and that common rights exist across groups. The two-tiered interest-based approach seeks to improve upon each of these theories by providing a clear means of specifying the rights of animals that is attuned to the importance of context but avoids essentializing group-based distinctions.

12

Welfare, Rights, and Non-ideal Theory

ROBERT GARNER

As the contributions to this book illustrate, the debate about the moral value of animal lives is complex, contentious, and not about to be resolved any time soon. According a right to life to at least some nonhuman animals has been the distinguishing characteristic of what has traditionally been taken as the "animal rights position" in the animal ethics debate (what I describe as the "species-egalitarian position"). Whatever the position adopted, this debate is one that is conducted within the realm of ideal theory. Ideal theories focus on the validity of a theory of justice or morality (the two are here used interchangeably) in relation to how far it is considered to approximate to the truth, insofar as normative arguments can arrive at such a determinate answer. But ideal theorizing, while essential, is not the only component of an adequate theory of justice, whether the focus is on humans or animals. Such a theory of justice must also engage with the terrain of non-ideal theory. That is, theories of justice must be judged in relation to their feasibility, or how far are they practically possible to achieve at any point in time, and how do we get from where we are now to where we want to be?

It is the aim of this paper to examine the issue of the moral value of animal lives in the context of non-ideal theory. Animal ethicists (and indeed animal advocates, too) have not engaged very rigorously with political strategy. The former, in particular, are, by and large, content to champion a particular normative theory as an ideal, irrespective of how far it departs from existing practice, and irrespective of the social, economic, and political constraints that may hinder, or even prohibit, its achievement. This paper seeks to engage with such political questions.

A number of characteristics of an effective non-ideal theory are identified. One such characteristic is the notion of finding a reasonable balance between morally divergent positions. One of the problems with this approach to non-ideal theory is that it appears to assume that the "reasonable balance" is an endpoint, that the inevitability of moral pluralism will prevent the eventual achievement of the chosen ideal theory. By contrast, following John Rawls, it is argued in this paper that non-ideal theory should be judged primarily by the degree to which it can facilitate, or is not inconsistent with, the ideal theory endpoint. The main aim of a non-ideal theory, therefore, should be to identify the most grievous example(s) of injustice or moral wrong-doing. Any valid non-ideal theory must prioritize the eradication of these injustices, and certainly should not be seen to hinder this goal.

The underlying assumption being made in the contrast between ideal and non-ideal theory is that a valid ideal theory of justice for animals might attach considerable value to animal lives, and might accord to animals a right to life. In actual fact, I dispute this assertion, arguing that it fails to take into account the ethical importance of the characteristics (wrapped up in the concept of personhood) possessed by most humans but not by most animals. Whatever the merits of this assertion, I would argue further that a valid non-ideal theory of justice for animals should not incorporate a right to life for nonhumans. Such a step is incompatible with the development of a realizable non-ideal theory of justice for animals, since the killing of animals for human gain plays a central role in all human societies.

This does not mean, however, that a valid non-ideal theory must accept the limitations of an animal welfare ethic. Such an ethic does not represent an optimum non-ideal theory. This is primarily because it is not morally permissible, in the Rawlsian sense, since it justifies inflicting suffering on animals provided that a significant benefit to humans accrues. This is to visit on animals a clear and fundamental injustice, thereby infringing Rawls's principle that a non-ideal theory should focus on the most urgent injustices. The infliction of suffering on animals represents such a major injustice because it can be established that animals have a right not to have suffering inflicted on them with greater certainty than the establishment of a right against them being killed.

There is an alternative to the acceptance of either the species-egalitarian version of animal rights or the animal welfare ethic. This is what I have described as the "sentience position," a model of animal rights that is recommended as the most appropriate non-ideal theory of justice or morality for animals. This is a theory of animal rights shorn of any claims about the moral value of animal lives, but which still holds that animals have a right not to suffer at the hands of humans. By contrasting the sentience position with the animal welfare ethic, it will be shown that the practical implications of the former are far-reaching and thereby morally permissible, and also politically realistic, partly at least because they do not require us to get involved in the thorny moral issue of killing animals.

12.1. IDEAL THEORY AND THE MORAL VALUE OF ANIMAL LIVES

The contributors to the first two sections of this book have grappled with the philosophical complexities of the value of animal lives, and with the extent to which it is a moral wrong to kill them. The importance of these questions has been heightened by the emergence, in the last forty years or so, of a reinvigorated animal rights movement, on both sides of the Atlantic, and a body of academic work that has challenged the dominant ethic that animal lives are of no moral import. It is not the purpose of this paper to provide a comprehensive account of this body of thought, and still less to engage in a critical analysis of it. Rather, the purpose is to provide a flavor of it in order to illustrate both its intractable and ideal character.

Of course, up until the nineteenth century, it was not just that conventional morality held that killing animals was of no moral import, but that we owed no direct duties to them at all. This position, echoed in the philosophy of thinkers such as Kant and Descartes, was transformed by a recognition, fuelled by the influence of utilitarianism, that the capacity of animals to suffer pain and pleasure was morally important. The contemporary moral orthodoxy is based around this recognition. Thus, the dominant animal welfare ethic holds that the infliction of suffering on animals is only morally permissible when it can be shown to be necessary, in terms of providing a substantial benefit to humans and/or other animals. The issue of death is, by itself, of little relevance to the animal welfare ethic. Rather, it is the degree to which animals suffer that is morally relevant.

The inferior moral status of animals enshrined in the animal welfare ethic has come under a sustained challenge in the past forty years or so. There is, of course, a utilitarian challenge to the moral orthodoxy associated, in particular, with Peter Singer (1990). The central characteristic of Singer's position is the claim that we should not presume that a human's interest in avoiding suffering should always be regarded as morally of greater weight than an animal's interest. As other papers in this book have revealed, the moral significance of death and killing in Singer's theory is a matter of much debate. On the issue of valuing animal lives, a greater contrast with the animal welfare ethic comes in the shape of the animal rights ethic associated, above all, with Tom Regan (1984).

Regan argues that at least some animals (to be precise, mammals one year of age and over) are what he calls "subjects of a life," and all human and nonhuman subjects-of-a-life have inherent value. This equal inherent value translates into strong moral rights. In particular, subjects-of-a-life with inherent value must be treated with respect as ends and not as means to an end. Regan, therefore, posits only one fundamental right, the right to respectful treatment, which derives from the inherent value of both humans and nonhuman animals. As a result, all uses of animals, irrespective of what is done to them while they are being used, become morally illegitimate. It goes without saying, too, that, according to Regan's version of animal rights—what might be described as the species-egalitarian version of animal rights—to kill animals for human purposes also infringes their right to respectful treatment. Thus, for Regan, the animal rights movement "is abolitionist in its aspirations. It seeks not to reform how animals are exploited . . . but to abolish their exploitation. To end it. Completely" (Cohen and Regan 2001, 127).

As earlier papers in this book have shown, the debate about the value of animal lives has tended to center on judgments about the harms incurred by humans and animals as a result of death. In particular, the critics of animal rights contend, humans are rational, self-conscious, autonomous persons, able to communicate in a sophisticated way and to act as moral agents. This is often converted into a shorthand claim that humans are persons while animals are not. The fact that most animals lack the characteristics of personhood challenges the claim that they have equivalent levels of interest in life and liberty to "normal" humans. A representative example of this approach is provided by Bonnie Steinbock, who states, "We do not subject animals to different moral treatment simply because they have fur and feathers, but because they are in fact different from human beings in ways that could be morally relevant" (1978, 247).

If the arguments of Steinbock and others are correct, then the animal welfare ethic may constitute a valid ideal theory of justice or morality for animals. That is, the animal welfare ethic argues that animals are morally inferior to humans (in the sense that they have less moral status because the interest they have, in avoiding suffering, counts for less than the interest humans have in avoiding suffering). Animals are therefore due only protection from unnecessary suffering according to the animal welfare ethic.

A critique of animal welfare from the perspective of ideal theory would, of course question the notion that animals are morally inferior to humans.

There are various different versions of this critique. One is to deny that all animals lack personhood. Another would be to deny the moral importance of personhood. It could be argued, for instance, that the animal welfare ethic accords too much moral weight to the fact that humans are persons and most animals are not. One, in my view illegitimate, version of this is to argue that little more than sentience is required for the attachment of considerable moral status, a position adopted by Gary Francione (2008). Another version is the argument that, while we can readily concede that the possession of personhood by humans may justify the claim that they have a greater interest than most animals in life and liberty, this does not justify the claim that an animal's interest in avoiding suffering is any less important to it than a human's. As a result, if humans have an interest in not having suffering inflicted on them by others, then, all things being equal, so do animals. As I will show below, this latter version is the basis of my preferred ideal and non-ideal theory of justice or morality for animals.

In the face of the strength of the claim that the possession of personhood is morally important, and that most animals would seem not to possess the characteristics of personhood, animal rights advocates have invoked the so-called *argument from marginal cases* (AMC) (Dombrowski 1997). The impeccable logic enshrined in the AMC is that if the possession of personhood is the key morally distinguishing characteristic, then those "marginal" humans (infants, the congenitally mentally disabled, and those with severe senile dementia) who are not persons either should be accorded the same moral status as animals. In short, if animals are deprived of a right to life, then so should marginal humans, and since we would not countenance such a demotion in the moral status of marginal humans, there is no justification for doing so in the case of animals.

As pointed out above, it is not the purpose of this paper to provide a detailed account of the value of animal lives in the modern animal ethics debate. What is important to note is the kind of debate it is. The flavor of it given above should be enough to reveal that it is heavily permeated by the rationalism typical of contemporary Western philosophy. It is about logical consistency, thoroughness, and precision. What it lacks is a sense of social and political realism, so that little attention is paid to the practical realization of the principles enunciated. Of course, this is particularly the case with the animal rights position associated with Regan. The prohibition on using and killing animals is so far removed from contemporary reality throughout the world that it is clearly, at some level at least, an inadequate theory.

It might be thought that, since they are, for the most part, defending the status quo, critics of the animal rights position are less prone to abstract theorizing, with little or no recognition of its applicability. This is not, however, always the case. Take the argument of the philosopher R. G. Frey. He suggests that the force of the AMC means that consistency demands that we either reject animal experimentation or "condone experiments on humans whose quality of life is exceeded by or equal to that of animals." Since it would be a mistake to forgo the benefits of animal experimentation, we should choose the latter. This choice is not made "with great glee, and rejoicing, and with great reluctance," but because "I cannot think of anything at all compelling that cedes all human life of any quality of greater value than animal life of any quality" (Frey 1983, 115).

Whatever its merits from the perspective of analytical moral philosophy, it is without doubt the case that it remains counter-intuitive to equate marginal humans with animals. Even though the AMC might be correct in theory, then, its lack of support in society means that it does not pass the "ought implies can" test. That is, humans will simply not allow the marginal members of their species to be regarded as morally inferior to other humans such that they are treated in the same way that animals are currently treated. As Evelyn Sztybel points out, "it is a fact of our contemporary, human nature that normal humans *cannot* pan-fry, say, mentally challenged humans." So, "if we *cannot* treat marginal humans on a par with nonhumans, let it not be said that we *ought* to do so" (2000, 340). Of course, it might be argued in reply that, as indicated above, it is precisely this response that animal rights advocates rely upon. That is, we would not treat marginal humans in the way that we currently treat animals, and therefore we should not treat animals with similar psychological characteristics in the same way. This is, however, to miss the point being made here, which is not a technical philosophical one but an empirical one to the effect that humans do not, by and large, regard animals as on a par morally with marginal humans, and would not accept them being so regarded.

12.2. THE CHARACTER OF NON-IDEAL THEORY

A solution to the discrepancy between the abstract theorizing of animal ethicists and the reality of animal use and killing in the modern world might lie in the application of non-ideal theory. In contemporary political discourse, the concept of non-ideal theory emerged as a result of increasing frustration by many at the discrepancy perceived between the abstract normative work of political philosophers, in which ideal political and moral principles are advocated, and the difficulty of applying such principles in the non-ideal real world. Rawls's theory of justice is often taken to be the classical example of an ideal theory. As he writes, "the nature and aims of a perfectly just society is the fundamental part of the theory of justice" (1971, 9).

Although use of the distinction between ideal and non-ideal theory dates back as far as Plato (Ypi 2010), the starting point for much of the contemporary discussion is Rawls. Although it is not compulsory to adopt his version, he does offer us a very useful framework that can be utilized to consider ideal and non-ideal theories relating to animals. Ideal theory, for Rawls, "presents a conception of a just society that we are to achieve if we can" (1971, 246). It is important to note, though, that ideal theory, for Rawls, should not be regarded as the same as utopianism, in the sense that it would be impossible to achieve. Even ideal theory must take into account the constraints afforded by human nature so that it probes "the limits of practicable political possibility" and "depicts an achievable social world" (Rawls 2001, 4, 6). Ideal theory, then, is equivalent to what Rawls describes as a "realistic utopia" which involves "taking men as they are and laws as they might be" (2001, 7).

David Miller is right when he writes that "even the basic concepts and principles of political theory are fact-dependent: their validity depends on the truth of some general empirical propositions about human being and human societies" (2008, 31). Here, we need to distinguish between "universal features of the human condition," which are unalterable, and "facts about particular societies, or types of society, and their inhabitants," which may not be (39). Determining whether a particular set of principles represents a realistic utopia involves, to some extent, relying, as Rawls points out, "on conjecture and speculation" where we have to argue "as best we can that the social world we envision is feasible and might actually exist, if not now then at some future time under happier circumstances" (2001, 12). Any theory which contains principles that are contrary to such "universal features of the human condition" would constitute a utopian, rather than an ideal, theory. What ideal theory does assume though, for Rawls, is "strict compliance," in the sense that it is assumed that not only are the laws of a society just, but also that "(nearly) everyone strictly complies with, and so abides by, the principles of justice" (13).

By contrast, non-ideal theory recognizes that an ideal theory may not be achievable, at least in the short term. For Rawls, non-ideal theory should be very much a secondary concern of political philosophers, who ought to focus on developing ideal theories. Others argue that Rawls's ideal theory, and that of many other political and moral philosophers, is so far removed from reality that ideal theories of justice ought to be given a much reduced status, if not dispensed with entirely. Colin Farrelly, for instance, argues that taking into account the political, social and economic realities within which ideal theory has to operate, "will mean that there is less room for armchair theorizing and that the primary focus will not be on winning a philosophical debate among first-order theories of justice" (2007b, 860; see also Sen 2009). Those who advocate this strong version of non-ideal theory are not then claiming simply that political pragmatism should prevail over normative political philosophy, but rather that any political or moral philosophy which does not take account of the non-ideal world in which it is attempting to influence and address is *normatively* deficient (see Dunn 1990; Carens 2000; Farrelly 2007a, 2007b; Sher 1997).

I do not share this hostility to ideal theory, at least if we adopt Rawls's version of it. It seems to me that, even if an ideal theory is not concerned with offering any desirable and achievable recommendations that are possible immediately, we still need it as a guide to the validity of any particular non-ideal theory. As Zofia Stemplowska rightly points out, "identifying the full extent of actual injustice requires knowledge of all that is wrong with the society in question, and this, in turn, requires knowledge of what a society in which no such wrongs were present would be like" (2008, 230). What *is* being claimed here is that a valid analysis of a particular theory of justice or morality must take into account the degree to which it is realizable in practice now, or in the medium or long term. It must not, that is, be divorced from questions relating to non-ideal constraints, whether they concern unsympathetic social, economic, or historical circumstances; moral disagreement; or human nature. This boils down to the well-known moral principle that "ought implies can." As Farrelly points out, "there is some conceptual incoherence involved in saying 'This is what justice involves, but there is no way it could be implemented'" (2007, 845).

One way in which non-ideal theory might be defined is in terms of the goal of finding a reasonable balance between morally divergent positions. A non-ideal theory would, in Farrelly's words, attempt to determine "what would constitute a *reasonable balance* between conflicting fundamental values" (2007b, 859). One of the problems with this approach to non-ideal theory, however, is that it appears to assume that the "reasonable balance" is an endpoint, that the inevitability of moral pluralism will prevent the eventual achievement of the chosen ideal theory. This is close to being a counsel of despair, albeit one that is perhaps politically realistic. By contrast, Rawls himself is very clear that non-ideal theory must be judged by the degree to which it facilitates, or is not inconsistent with, the ideal theory end point. For Rawls, then, the goal of non-ideal theory is to consider how the long-term goal of ideal theory "might be achieved, or worked toward, usually in gradual steps" (1971, 246).

Rawls provides some guidance to judging the effectiveness of a non-ideal theory. It should, he argues, look "for courses of action that are morally permissible and politically possible as well as likely to be effective" (2001, 89). The meaning of the second of these is well understood, albeit perhaps difficult to determine. The moral permissibility of a course of action, for Rawls, is a function of the degree to which it removes the most grievous or most urgent injustice, the one that departs the most from the ideal theory (1971). Finally, Rawls holds to the view that the effectiveness of a non-ideal theory can be judged by the degree to which it moves society toward the ideal position. This definition of effectiveness is useful because it enables us to distinguish non-ideal theory from the common-sense, "second best," position that if we cannot achieve all that justice demands we should, instead, get what we can (Simmons 2010, 25). Rather, Rawls insists that only those courses of action that are functional for the achievement of the ideal theory are permissible. Thus he is committed to the position that even where it is politically possible and morally permissible to remove an injustice, we should only do so if it does not impede the process whereby our ideally just endpoint is achieved.

12.3. NON-IDEAL THEORY AND A RIGHT TO LIFE FOR ANIMALS

What I have argued so far is that a valid theory of justice must consider what is the most effective route to the establishment of the principles enshrined in the ideal theory. It must, that is, include non-ideal theory. The use of non-ideal theorizing in the animal ethics debate in general, and in the debate about the value of animal lives in particular, allows us to factor in questions of feasibility and strategy.

It should have become obvious by now that I do not accept the validity of including the granting of a right to life to animals as a component of an adequate non-ideal theory. Such a provision, by definition, cannot be part of a non-ideal theory of justice or morality for animals, because it automatically prohibits the institutional uses of animals—as sources of food and (to a large degree) as scientific subjects. Such a prohibition would, at a stroke, achieve the chief objective of the animal rights movement, and would represent a massive, and some would say inconceivable, departure from current practice. It therefore represents an endpoint, which is the terrain of ideal theory.[1]

Of course, the species-egalitarian version of animal rights, including a prohibition on the killing of animals, might still serve as a valid ideal theory of justice or morality for animals. It is not the prime purpose of this paper to assess this claim, although I have argued elsewhere (Garner 2013) that the species-egalitarian version of animal rights ought to be rejected as an appropriate *ideal* theory. This is the case, I would argue, partly because it is mistaken on the grounds of ethical principle. That is, for some of the reasons expressed above, I do not think that granting a right to life to animals is a valid component of an ideal theory of justice or morality. I am not suggesting that animal lives have no moral value, but to regard them as equally morally valid to those of humans is to ignore the characteristics (of personhood) that humans possess and (most) animals do not. I reject, in other words, the species-egalitarian version of animal rights associated with Regan, because it fails to take into account the moral significance of those interests—in liberty and in life—associated with persons, as well as the fact that persons have a greater interest in life and liberty than nonpersons.

This ethical critique of the species-egalitarian version of animal rights leads me to recommend an alternative ideal theory of justice for animals. This I describe as the "enhanced sentience position." This position essentially gives Singer's equal consideration of interests principle a rights-based tweak. Deriving from Cochrane (2012) and Rachels (1990), among others, it suggests that, while animals may not have a right to life or liberty, this does not mean that they cannot possess a right not to suffer at the hands of humans. The crucial insight here is that adopting an animal rights position does not require us to justify moral equality between humans and animals. Instead, the like interests of humans and animals are to be treated equally, and their unlike interests are to be treated unequally. Differentiating morally between the interests of humans and animals allows us, then, to paint a much more complex, and realistic, picture of our moral obligations to animals than the species egalitarianism of the dominant strand of animal rights thinking.

The main contours of the enhanced sentience position can be best explained by considering what is wrong with animal welfare as an ethic. The animal welfare ethic is right to draw attention to the fact that humans and animals differ in ways that are morally relevant. I broadly agree with what has become the moral consensus that normal adult humans possess a greater interest in life and liberty than most animals, and that this ought to be reflected in a calculation of the respective moral importance of humans and animals. However, advocates of the animal welfare ethic then, it seem to me, take a wrong turn. They conclude from this that *all* human interests are more important morally than *all* animal interests. In other words, they assume that because most humans have a greater interest in liberty and continued life than most animals that they have a greater interest in other things, too. But this does not follow. Indeed, all things being equal, it is difficult to avoid the conclusion that an animal's interest in avoiding suffering is equivalent to a human's interest in avoiding suffering.

The fact that some animal interests are equivalent in strength to those of humans is not strong enough to support the species-egalitarian version of animal rights. It is a valid claim that suffering of the same intensity and duration has the same moral weight whether endured by humans or animals. It is also a valid claim that the interest that animals have in not suffering can be translated into a right not to suffer at the hands of humans, so that if humans have such a right then so do animals. The same cannot, however, be claimed in the case of the moral value of animal lives. Even though animals may have some interest in continued life, it would seem inappropriate to accord to them a right to life, because human lives are morally more important and, in the event of a conflict, must take precedence over those of animals.

The enhanced sentience position would impose considerable restrictions on what we are permitted to do to animals. In the first place, the moral value of animal lives would have to be factored into our ethical judgments. It would mean that only very significant human benefits (such as the protection of human lives) would have to accrue from the loss of animal lives at human hands. This would allow for the use (and killing) of animals when human lives are at stake, such as in scientific procedures where it could be shown that their use was essential for the development of life-saving medical treatments. However, the enhanced sentience position would be even more restrictive than this suggests. This is because the according to animals of a right not to suffer would mean that it would be illegitimate to inflict suffering on animals *even if* very significant human benefits, such as the saving of human lives, would accrue.

I have argued that according to animals a right to life should not be part of a non-ideal theory of justice for animals. What I call the sentience position serves this purpose by removing any reference to the value of animal lives, including the right to life stricture. The sentience position does prohibit morally the infliction of suffering on animals for human benefits, but at the same time accepts that humans can still, under certain circumstances, use them. Because it does not engage at all with the question of the value of animal lives, sacrificing animal lives for human benefit is not regarded as problematic ethically.

The sentience position is far from being a defense of the status quo. Indeed, it is very demanding on human beings, because it rules out the infliction of suffering on animals for our benefit. On the other hand, it is more realistic than a theory based on denying the ethical validity of using animals as, for example, sources of food and as experimental sub-jects irrespective of what is done to them while they are being used (the species-egalitarian version of animal rights). Likewise, it is more realistic than the enhanced sentience position, which accepts that animals do have an interest in continued life, albeit not as great an interest as that pos-sessed by humans.

12.4. DEFENDING THE SENTIENCE POSITION

The sentience position, I would argue, is the most appropriate non-ideal theory of justice for animals. To see that this is so, it should be noted firstly that it meets Rawls's criteria for a valid non-ideal theory. As we saw above, Rawls argues that non-ideal theory should be effective, morally permis-sible, and politically achievable. The starting point for Rawls is to identify the most grievous examples of injustice. That is, Rawls argues that it is only morally permissible to prioritize the most urgent cases of injustice, the cases which diverge furthest from our ideal theory. Rawls's lexical ordering of his major principles of justice makes it easy for him to specify what con-stitutes the most serious cases of injustice. Thus, violations of his liberty principle (requiring extensive and equal basic liberties) are more serious than infringements of the "equal opportunity principle" or the "difference principle."

What is interesting here is that Rawls assumes, correctly of course, that humans are granted bodily integrity, so that it would be unjust to kill or maim members of our own species, a position accepted universally in the case of the developed countries his theory of justice is meant to apply to. No such protection is granted to animals, which, of course, regularly suf-fer at the hands of humans and are killed by them for a variety of reasons. The enhanced sentience position, the goal to which it is argued we should be heading, does not stipulate that animals have a right to life or liberty, although it does place some value on animal lives. What it does insist upon, however, is that animals have a right not to suffer, irrespective of the ben-efits that might accrue to humans as a result. It would seem appropriate, therefore, to regard eliminating suffering as the most urgent injustice in the case of animals.

In the context of non-ideal theory, therefore, the position I have described as the "sentience position" reflects most accurately the urgent need to eliminate animal suffering at the hands of humans. It therefore has a distinct advantage over the animal welfare ethic as a viable non-ideal theory. To see that this is so, we need to delineate the key differences between the animal welfare ethic and the sentience position. According to the animal welfare ethic, it is permissible, morally, to inflict suffering on an animal, provided that the benefits to be gained from so doing are perceived to be sufficiently large. The sentience position, on the other hand, rules out such a cost-benefit approach. *Whatever* the benefit that might accrue to humans, or other animals for that matter, practices that inflict suffering on animals are prohibited.

The contrast between the sentience position and the animal welfare ethic reveals that the moral inferiority of animals postulated in the latter can justify a great deal of animal exploitation and, more specifically, suffering. Such a position, therefore, is not, in practice, a reasonable balance between divergent moral positions. Moreover, by allowing the infliction of suffering, provided that a significant benefit to humans accrues, the animal welfare ethic visits on animals a clear and fundamental injustice or moral wrong, thereby infringing Rawls's principle that a non-ideal theory should focus on the most urgent injustices. By contrast, an ethic based on the alleviation of animal suffering at the hands of humans has a greater claim to be the ideal non-ideal theory for animals.

It might be argued at this point that the sentience position falls foul of another of Rawls's criteria for valid non-ideal theory. This is that the sentience position is not politically possible. I do not have the space here to investigate this fully. All I will say is that it is a distinct possibility that the degree to which popular opinion is prepared to recognize a higher moral status for animals than the animal welfare ethic allows for can be underestimated, not least because, in the public mind, the animal welfare ethic is often conflated with the, often far-reaching, conclusions of animal welfare science. It is also important, I would argue, for an animal ethic to be aspirational and to be readily distinguishable from an animal welfarism that has, in practice at least, lost its meaning because of its ubiquity.

12.5. CONCLUSION

As a final point, and remembering the theme of this volume, the key point is that it is not necessary to get involved in interminable debates about the moral value of animal lives, and how far it is justified to kill animals for human benefit, in order to challenge the current ways in which animals are treated. There is a gap, ideal and non-ideal, between the species-egalitarian version of animal rights and the animal welfare ethic. Focusing on animals' right not to have suffering inflicted on them takes us a long way down the traditional animal rights road without having to transform the traditional, and solidly entrenched, consensus that human lives are of greater value than those of animals.

NOTE

1. Strictly speaking, prohibiting the killing of animals is not the desired endpoint if one regards the use of animals as morally illegitimate, since not all uses of animals result in their deaths (the keeping of companion animals being the obvious example). This "use position" is the logical consequence of accepting the position, · derived from Regan, that to use animals is to infringe their right to be treated with respect.

Afterword

PETER SINGER

Forty years ago, in the first chapter of *Animal Liberation* (1975), I argued that "speciesism" is a prejudice or bias, in some ways akin to racism and sexism, that leads us to give less weight to the suffering of nonhuman animals than we give to the similar, or sometimes lesser, suffering of members of our own species. I then wrote:

> So far I have said a lot about the infliction of suffering on animals, but nothing about killing them. This omission has been deliberate. The application of the principle of equality to the infliction of suffering is, in theory at least, fairly straightforward.... The wrongness of killing a being is more complicated. (Singer 2009, 17)

The complication arises, I explained, from the fact that even if we focus only on human beings, people hold different views about why killing is wrong, as controversies over abortion and euthanasia indicate. What lies behind these different views? To some extent, the issue is whether one accepts the idea of "the sanctity of life." Opponents of abortion and euthanasia often use this phrase, although typically they mean by it "the sanctity of human life." Only a tiny minority of those who oppose abortion and euthanasia also oppose the killing of animals for food, although there are nutritionally adequate alternatives to eating meat. Similarly, people who would be horrified at the idea of killing a human being in order to advance medical research—even if the human is an infant born without a cortex, and thus incapable of feeling pain—are generally willing to accept killing sentient nonhuman animals for the same goal. The most obvious explanation for these differences is speciesism. (Religious beliefs play a role, but many religious beliefs are themselves speciesist, such as the belief that all and only human beings are made in the image of God, or have immortal souls.)

On the other hand, rejecting speciesism does not, in itself, entail that we should equate killing a dog or a pig or a chicken with killing a normal human being. It is plausible to hold that the wrongness of killing a being depends on the quality, and not the sanctity, of its life. This quality will vary, not only with its circumstances, but also with the kind of mental life the being has—for example, its capacities to enjoy its life, or its ability to plan its future. Whatever features we choose, however, we will have to accept that there are some nonhuman animals that have these features to a greater extent than some human beings. For example, a dog will be better able to enjoy its life than an anencephalic human infant born with only a brainstem and no cortex. Hence any line or lines we draw indicating that taking the life of one kind of being is, because of that being's quality of life, more serious than taking the life of a different kind of being will not be identical to the boundary between human beings and nonhuman animals.

I wrote *Animal Liberation* for a broad audience, because I wanted to persuade as many people as possible that the way we treat animals is indefensible and in need of radical change. To go deeply into the question of when killing animals is wrong would have been incompatible with the aim of attracting the widest possible readership. Hence, after a brief discussion along the lines just outlined, I put the topic of killing animals aside. The papers in this volume offer ample evidence that I was right to do so. Forming a consistent and intuitively appealing view about when killing is wrong is an extraordinarily difficult task.

The fact that making up our minds about the killing of animals is difficult does not, of course, mean that it is unimportant. Here is one way of assessing its importance. Many people think that abortion is an important moral issue. The World Health Organization estimates that 40–50 million abortions are carried out each year. As Tatjana Višak and Robert Garner point out in their introduction to this book, the UN Food and Agriculture Organization (FAO) conservatively estimates that 65 billion land animals are killed for food each year. That's more than a thousand times as many animals killed for food as fetuses killed by abortions. If we add to the 65 billion the vast number of fish killed for food (estimated to be between 1 and 3 trillion), then the number of vertebrate animals killed for food is at least 20,000 times greater than the number of fetuses killed. So the need to determine whether it is wrong to kill nonhuman animals is at least as pressing as the need to determine whether it is wrong to kill a human fetus, for if the killing of nonhuman animals is generally wrong, then it is wrong-doing that is happening on a vast scale.

After avoiding lengthy discussion of the issue of killing animals in *Animal Liberation*, I took it up four years later, in *Practical Ethics* (1979), and continued to grapple with it in subsequent editions of that book, in 1993 and 2011. In this volume, Christopher Belshaw, Tatjana Višak, and Shelly Kagan all refer to my support of the controversial idea of *replaceability*; that is, the idea that in some circumstances one can defend the killing of animals on the grounds that if they are killed, they will be replaced by others living good lives. This idea was not invented by academic philosophers; it goes back at least as far as the nineteenth-century British essayist Leslie Stephen (perhaps most famous now as the father of the novelist Virginia Woolf), who wrote: "Of all the arguments for Vegetarianism none is so weak as the argument from humanity. The pig has a stronger interest than anyone in the demand for bacon. If all the world were Jewish, there would be no pigs at all" (Stephen 1896). If pigs have lives worth living, Stephen is suggesting, that is better than there being no pigs at all, and a necessary condition for the existence of large numbers of pigs is that they are killed for food, because otherwise no one would go to the expense and trouble of raising them. This argument keeps popping up, and it has been used by many other defenders of meat eating, including Michael Pollan (2006, 310), in his best-selling *The Omnivore's Dilemma*.

In the first edition of *Animal Liberation* I rejected this defense of meat eating, citing Henry Salt, a contemporary of Stephen and an early advocate of animal rights, who responded to Stephen's argument by saying, "A person who is already in existence may feel that he would rather have lived than not, but he must first have the terra firma of existence to argue from; the moment he begins to argue as if from the abyss of the non-existent, he talks nonsense, by predicating good or evil, happiness or unhappiness, of that of which we can predicate nothing" (Salt 1914). This is essentially the position taken by Tatjana Višak when she holds that "existing as opposed to never existing cannot benefit or harm an animal" and defends this view by arguing that we cannot compare a state of affairs in which an animal exists with a state of affairs in which the animal does not exist in terms of this animal's welfare. This is impossible, according to Višak, because an animal has no lifetime welfare level in a world in which it does not exist.

Logical as this objection to Stephen's defense of meat eating may seem, between first writing *Animal Liberation* and *Practical Ethics* I came to the view that it is not tenable. This was, in part, because of the problem of the asymmetry between bringing a happy child into existence and bringing a miserable child into existence, a problem discussed in depth in this volume by McMahan and Kagan as well as Višak. Although most people feel that there is no obligation to bring a child into existence merely because that child is likely to live a good life, almost everyone agrees that it would be wrong to bring into existence a child who, perhaps because of a genetic condition, would live a life of unmitigated suffering and then die. Višak bites the bullet on this implication of her view, while trying to argue that its implications are less disturbing than one might at first assume. On reflection, it seems to me more plausible to accept the other horn of the dilemma: the fact that an as-yet-unconceived child is likely to have a happy life is, other things being equal, a reason for bringing that child into existence.

One further problem with person-affecting views should also be mentioned. If we hold that life for human beings is, on balance, positive and has the potential to get much better still as our growing technological capacities make it easier for us to give everyone a good life, this gives us grounds for thinking that the extinction of our species would be one of the greatest possible tragedies. After all, if we do not become extinct, billions of humans may flourish for billions of years, making both moral and scientific progress, and living rich, fulfilling lives at a level beyond our imagination. On the person-affecting view, however, extinction is a tragedy only if it harms existing humans, or those who would exist anyway. There is some evidence that childless couples are, on average, happier than those who have children. Suppose that this evidence leads all existing fertile humans to believe that they would be better off if they were to have themselves sterilized. It would not, on a person-affecting view, be wrong for them to do so. I hold, on the contrary, that to deliberately eliminate the prospect of billions of humans living rich, happy lives, merely so that one's own life will go slightly better, would be to act very wrongly indeed (see Parfit 1984, 453–4; Bostrom 2013, 15–31).

The view I have taken on these questions implies that we can compare existence with non-existence. There are independent grounds for thinking that this is possible. Consider the following three scenarios:

Non-existence: My parents decide not to have children, so I never exist.

Non-conscious life: My parents conceive a fetus, genetically identical with me, but before the fetus develops consciousness, it suffers a massive brain hemorrhage. The resulting child is born in a vegetative state and remains that way, never gaining consciousness until, eighty years later, it dies.

A happy life: My parents conceive a child (me) and all goes well. I live happily until eighty, when I die.

Obviously, I prefer a happy life to a non-conscious life. As regards my own welfare, I am indifferent between non-existence and a non-conscious life (the birth of a child that will never gain consciousness has many negative side effects for the child's parents, and others, but we are putting them aside here). If I prefer a happy life to a non-conscious life, and am indifferent between a non-conscious life and non-existence, it seems that I also prefer a happy life to non-existence. A happy life has positive value, and a life of unrelieved misery has negative value, but a life without consciousness has zero value. So does non-existence.

In giving zero welfare to a state of non-existence as well as to a non-conscious existence, I am agreeing with Nils Holtug. Višak describes this as a "common mistake," but her explanation of the mistake leaves it mysterious as to how we can make judgments that we don't seem to have any logical (as opposed to emotional or empirical) difficulty in making. For example, many couples who know that they carry a genetic defect have counseling to help them decide whether to have their own biological child. Some of that counseling will be about how a child with a genetic defect will affect their own lives, and the lives of others close to them, but one important factor is likely to be whether their child will have a life that is too miserable to be worth living. The problem is not just that, as we have seen, Višak has to bite the bullet on the moral judgment that there is nothing directly wrong with the parents conceiving a child whose life will be miserable. She must also hold that when we say that it would be bad for such a child to be brought into existence, we are making some kind of logical error. We are seeking to compare values that are incommensurable, and judging one state worse than another when "existing . . . as opposed to never existing . . . cannot be better or worse for an individual" (this volume, p. 119).

Višak takes the view that I can compare my lifetime welfare in the possible world in which I die tonight with my lifetime welfare in the world in which I live for many more years. No one questions the coherence of such judgments, although in this volume Steven Luper raises a related puzzle about how I can be better off at a time when I no longer exist. If, however, we can compare my lifetime welfare in the possible world in which I die tonight with that in which I live for many more years, then it must be possible to make that comparison for someone much younger than me who also dies tonight instead of living for many more years. How much younger? Can we compare the lifetime welfare of an infant who lives for one day, enjoying suckling at her mother's breast, and then dies, with the lifetime welfare of the same infant who lives to eighty? What if the infant manages just to take one breath before dying? What if the infant is stillborn? What if the fetus experiences some dull conscious sensations and then dies in the womb? What if the fetus dies before it has any conscious experiences at all? I ask these questions to show that on a person-affecting view, it becomes critically important to answer the question of when a being comes into existence. The problem is that, whether what we consider crucial is the existence of the organism or the existence of the person, the line between existence and non-existence is not sharp enough to bear the weight of the supposedly profound difference between an act that ends the life of an existing being and an act that supposedly does no harm because no being ever existed to be made worse off than it would otherwise have been.

An impersonal view avoids this problem in a straightforward way. It does not have to say, with Višak, that existence and non-existence are incommensurable because one of these states cannot be better or worse for an individual than the other. On the impersonal view states do not have to be better or worse for someone. They can be better or worse impersonally, or to use Sidgwick's phrase, "from the point of view of the universe" (Sidgwick 1907, 382). This is true even if it is combined with preference utilitarianism, which is the position I defended in the first and second editions of *Practical Ethics* (1979 and 1993). Around the time when I was revising the third edition (2011) I became more doubtful about preference utilitarianism, and began to think of hedonistic utilitarianism as a more defensible option, but I was not yet ready to rewrite the book along those lines. Only in *The Point of View of the Universe* (Lazari-Radek & Singer 2014, chaps. 8, 9) did my coauthor and I argue in favor of hedonistic utilitarianism.

Let us assume, for simplicity, that hedonistic utilitarianism is sound. Then, if we reject the person-affecting view because of the difficulties we have discussed, impersonal hedonistic utilitarianism has a ready explanation for why it is good, other things being equal, to bring a happy child into existence, and wrong to bring a miserable child into existence. The happy child adds to the net surplus of pleasure in the world (if there is one; otherwise it reduces the net deficit), and the miserable child does the opposite. It is, on the impersonal view, obvious why not having a child, and having a permanently unconscious child, can both be rated at zero, considered apart from the side effects they have on others: intrinsically, neither contributes anything positive or negative to the sum total of utility. There is no problem in explaining the importance of the moment when a being comes into existence (whatever that moment may be), because it isn't important. Impersonal utilitarians want to maximize the total amount of utility in the world, and it doesn't matter, in itself, whether we do it by adding to the utility experienced by an existing being, or by bringing into existence a being who would not otherwise have existed at all.

Admittedly, hedonistic utilitarianism has ample counter-intuitive implications of its own. In *Practical Ethics* I tried to show that preference utilitarianism can distinguish between persons (that is, beings capable of understanding that they have a future, and of having desires about that future) and beings who lack this capacity and so are not persons. Persons, I argued, are not replaceable in the way that beings who are not persons are. This led to various complications, including the "debit model" of preference utilitarianism, which Kagan discusses in his paper in this volume. If we accept hedonistic utilitarianism rather than preference utilitarianism, we don't need the complexities of the debit model, and the distinction between persons and other conscious beings becomes less significant. Instead, replaceability becomes acceptable, at least in theory, for humans as well as for animals.

In practice, a capacity somewhat similar to that of knowing that one has a future remains relevant. It will normally be worse to kill beings who are capable of learning that others like them have been killed, and of fearing that they too will be killed, for this will make their lives worse. That is why Jeremy Bentham, protesting against the cruelty of inflicting the death penalty on mothers who kill their newborn infants, pointed out that this is an offense "of a nature not to give the slightest inquietude to the most timid imagination," for all those who come to learn of the offense are themselves too old to be threatened by it (Bentham 1840, 33). Something similar will apply to the killing of animals when this is done in such a way that it does not terrify or distress other animals, and this is a reason why the replaceability of animals, in the right circumstances, will be easier to defend than the replaceability of humans. The extent to which the death of a being will cause grief and loss to others who love that being is also obviously relevant. That is not going to satisfy many people, who will continue to object to the idea that humans could be replaceable, if none of these side effects apply. There is no consistent solution to the problems we are discussing that does not involve biting at least one bullet. If you are seeking to discover which bullet is the least unpalatable, the papers in this volume will have been a good beginning.

REFERENCES

Anderson, E. 2004. "Animal Rights and the Values of Nonhuman Life." In *Animal Rights: Current Debate and New Directions*, edited by M. Nussbaumand Sunstein, 277–98. New York: Oxford University Press.

Apel, K. 1980. "The a Priori of the Communication Community and the Foundations of Ethics: The Problem of a Rational Foundation of Ethics in the Scientific Age." In *Towards a Transformation of Philosophy*, 225–300. London: Routledge & Kegan Paul.

Appleby, M. C., and P. Sandøe. 2002. "Philosophical Debate on the Nature of Well-Being: Implications for Animal Welfare." *Animal Welfare* 11: 283–94.

Arras, J., and E. Fenton. 2009. "Bioethics and Human Rights: Access to Health-Related Goods." *Hastings Center Report* 29: 27–38.

Arrhenius, G. 2000. *Future Generations: A Challenge for Moral Theory*. Uppsala, Sweden: Uppsala University.

Arrhenius, G. 2009. "Can the Person-Affecting Restriction Solve the Problems in Population Ethics?" In *Harming Future Persons: Ethics, Genetics and the Nonidentity Problem*, edited by M. A. Roberts and D. T. Wasserman, 289–314. Dordrecht, The Netherlands: Springer.

Arrhenius, G. Forthcoming. *Population Ethics: The Challenge of Future Generations*. Oxford: Oxford University Press.

Arrhenius, G., and W. Rabinowicz. 2015. "The Value of Existence." In *The Oxford Handbook of Value Theory*, edited by I. Hirose and J. Olson, 424–43. Oxford: Oxford University Press.

AVMA (American Veterinary Medical Association). 2013. *AVMA Guidelines for the Euthanasia of Animals*. Schaumburg, IL: AVMA. https://www.avma.org/KB/Policies/Documents/euthanasia.pdf.

Bagnoli, C. 2013. *Constructivism in Ethics*. New York: Cambridge University Press.

Barron, A., R. Maleszka, P. Helliwell, and G. E. Robinson. 2009. "Effects of Cocaine on Honey Bee Dance Behavior." *Journal of Experimental Biology* 212: 163–68.

Barron, A., E. Søvik, and J. Cornish. 2010. "The Roles of Dopamine and Related Compounds in Reward-Seeking Behavior across Animal Phyla." *Frontiers in Behavioral Neuroscience* 4: 163.

Begon, M. 2003. "Disease: Health Effects on Humans, Population Effects on Rodents." In *Rats, Mice and People: Rodent Biology and Management*, edited by G. Singleton, L. Hinds, C. Krebs, and D. Spratt, 13–19. Canberra: Australian Centre for International Agricultural Research.

Belshaw, C. 2005. *10 Good Questions about Life and Death* Oxford: Blackwell.

Belshaw, C. 2009. *Annihilation: The Sense and Significance of Death*. Stocksfield, UK: Acumen.

Belshaw, C. 2012. "Death, Value, and Desire." In *The Oxford Handbook of Philosophy of Death*, edited by edited by B. Bradley, F. Feldman, and J. Johansson, 274–296. Oxford: Oxford University Press.

Benatar, D. 2006. *Better Never to Have Been: The Harm of Coming into Existence*. Oxford: Oxford University Press.

Bentham, J. 1840. *The Theory of Legislation*. Vol. 2. Translated from the French of Etienne Dumont by Richard Hildreth. Boston: Weeks, Jordan.

Bermúdez, J. 2007. *Thinking without Words*. Oxford: Oxford University Press.

Blackorby, C., and David Donaldson, 1991. "Normative Population Theory: A Comment." In *Social Choice and Welfare* 8: 261–267.

Bostrom, N. 2013. "Existential Risk Prevention as Global Priority." *Global Policy* 4: 15–31.

Bradley, B. 2004. "When Is Death Bad for the One Who Dies?" *Nous* 38: 1–28.

Bradley, B. 2009. *Well-Being and Death*. Oxford: Oxford University Press.

Bradley, B. 2013. "Asymmetries in Benefiting, Harming and Creating." *Journal of Ethics* 17: 37–49.

Bradley, B., F. Feldman, and J. Johansson, eds. 2013. *The Oxford Handbook of Philosophy of Death*. Oxford: Oxford University Press.

Bradley, B., and K. McDaniel. 2013. "Death and Desires." In *The Metaphysics and Ethics of Death*, edited by J. Taylor, 118–33. Oxford: Oxford University Press.

Brambell, F. W. R. 1965. *Report of the Technical Committee to Inquire into the Welfare of Animals Kept under Intensive Livestock Husbandry Systems*. London: H.M.S.O.

Brandt, R. 1979. *A Theory of the Good and the Right*. Oxford: Clarendon.

Broadie, A., and M. Pybus. 1974. "Kant's Treatment of Animals." *Philosophy* 49, no. 190: 375–83.

Broom, D. M. 2011. "A History of Animal Welfare Science." *Acta Bioetheoretica* 59: 121–37.

Broom, D. M., and K. G. Johnson. 1993. *Stress and Animal Welfare*. London: Chapman & Hall.

Broome, J. 1993. "Goodness is Reducible to Betterness: The Evil of Death Is the Value of Life." In *The Good and the Economical*, edited by P. Koslowski and Y. Shionoya. Berlin: Springer-Verlag, pp. 70–86.

Broome, J. 1999. *Ethics out of Economics*. Cambridge: Cambridge University Press.

Broome, J. 2004. *Weighing Lives*. Oxford: Oxford University Press.

Broome, J. 2013. "The Badness of Death and the Goodness of Life." In *The Oxford Handbook of Philosophy of Death*, edited by B. Bradley, F. Feldman, and J. Johansson. Oxford: Oxford University Press, pp. 218–233.

Bruijnis, M. R. N., F. L. B. Meijboom, and E. N. Stassen. 2013. "Longevity as an Animal Welfare Issue Applied to the Case of Foot Disorders in Dairy Cattle." *Journal of Agricultural and Environmental Ethics* 26: 191–205.

Buchanan, A. et al 2000. *From Chance to Choice: Genetics and Justice*. Cambridge: Cambridge University Press.

Bykvist, K. 2007. "The Benefits of Coming into Existence." *Philosophical Studies* 135: 335–62.

Caney, S. 2007. "Global Poverty and Human Rights: The Case for Positive Duties." In *Freedom from Poverty as a Human Right: Who Owes What to the Very Poor?*, edited by T. Pogge, 275–302. Oxford: Oxford University Press.

Carens, J. 2000. *Culture, Citizenship and Community: A Contextual Exploration of Justice and Evenhandedness*. Oxford: Oxford University Press.

Carruthers, P. 1992. *The Animals Issue*. Cambridge: Cambridge University Press.

Charleston, B., B. M. Bankowski, S. Gubbins, M. E. Chase-Topping, D. Schley, R. Howey, and M. E. J. Woolhouse. 2011. "Relationship between Clinical Signs and Transmission of an Infectious Disease and the Implications for Control." *Science* 332: 726–29.

Chittka, L., and L. Niven. 2009. "Are Bigger Brains Better?" *Current Biology* 19: 995–1008.

Cigman, R. 1981. "Death, Misfortune and Species Inequality." *Philosophy and Public Affairs* 10: 47–64.

Cochrane, A. 2012a. *Animal Rights without Liberation*. New York: Columbia University Press.

Cochrane, A. 2012b. "Evaluating 'Bioethical Approaches' to Human Rights." *Ethical Theory and Moral Practice* 15: 309–22.

Cochrane, A. 2013a. "Cosmozoopolis: The Case against Group-Differentiated Animal Rights." *Law, Ethics and Philosophy* 1: 127–41.

Cochrane, A. 2013b. "From Human Rights to Sentient Rights." *Critical Review of International Social and Political Philosophy* 16: 655–75.

Cohen C., and T. Regan. 2001. *The Animal Rights Debate*. Lanham, MD: Rowman & Littlefield.

Darwall, S. 2006. *The Second-Person Standpoint: Morality, Respect, and Accountability*. Cambridge, MA: Harvard University Press.

Darwall, S. 2007. "Reply to Korsgaard, Wallace, and Watson." *Ethics* 118, no. 1: 52–69.

Dasgupta, P. 1995. *An Inquiry into Well-Being and Destitution*. Oxford: Oxford University Press.

Dawkins, M. 2008. "The Science of Animal Suffering." *Ethology* 114: 937–45.

Dean, R. 2006. *The Value of Humanity in Kant's Moral Theory*. Oxford: Oxford University Press.

DeGrazia, D. 1996. *Taking Animals Seriously: Mental Life and Moral Status*. New York: Cambridge University Press.

DeGrazia, D., and A. Rowan. 1991. "Pain, Suffering, and Anxiety in Animals and Humans." *Theoretical Medicine* 12, no. 3: 193–211.

Denis, L. 2000. "Kant's Conception of Duties Regarding Animals: Reconstruction and Reconsideration." *History of Philosophy Quarterly* 17, no. 4: 405–23.

Dickens C., 1977. *Bleak House*. New York: W. W. Norton.

DiGiacomo, N., A. Arluke, and G. Patronek. 1998. "Surrendering Pets to Shelters: The Relinquisher's Perspective." *Anthrozoos* 11: 41–51.

Dombrowski, D. 1997. *Babies and Beasts: The Argument from Marginal Cases*. Chicago: University of Illinois Press.

Donaldson, S., and Kymlicka, W. 2011. *Zoopolis: A Political Theory of Animal Rights*. Oxford: Oxford University Press.

Dunn, J. 1990. "Reconceiving the Content and Character of Modern Political Community." In *Interpreting Political Responsibility*, edited by J. Dunn, 193–215. Cambridge: Polity Press.

Dworkin, R. 1984. "Rights as Trumps." In *Theories of Rights*, edited by J. Waldron, 153–67. Oxford: Oxford University Press.

Dworkin, R. 1993. *Life's Dominion* New York: Alfred A. Knopf.

Enoch, D. 2006. "Agency, Shmagency: Why Normativity Won't Come from What Is Constitutive of Action" *Philosophical Review* 115, no. 2: 169–98.

Epicurus 1926. "Letter to Menoeceus." In *The Extant Remains*, translated by C. Bailey. Oxford: Clarendon.

Epicurus 1964. *Letters, Principal Doctrines, and Vatican Sayings*. Translated by Russel Geer. Indianapolis, IN: Bobbs-Merrill.

European Commission. 2010. *Directive 2010/63/EU of the European Parliament and of the Council of 22 September 2010 and of the Council of 22 September 2010 on the Protection of Animals Used for Scientific Purposes*. Brussells: European Commission. http://eurlex.europa.eu/LexUriServ/LexUriServ.do?uri=OJ:L:20 10:276:0033:0079:eN:PDF.

Farrelly, C. 2007a. *Justice, Democracy and Reasonable Agreement*. Basingstoke, UK: Palgrave Macmillan.

Farrelly, C. 2007b. "Justice in Ideal Theory: A Refutation." *Political Studies* 55: 844–64.

FAWC (Farm Animal Welfare Council). 2009. *Farm Animal Welfare in Great Britain: Past, Present and Future*. London: FAWC.

Fehige, C. 1998. "A Pareto Principle for Possible People." In *Preferences*, edited by Christoph Fehige and Ulla Wessels. Berlin: Walter de Gruyter, pp. 508–541.

Feinberg, J. 1974. "The Rights of Animals and Future Generations." In *Philosophy and Environmental Crisis*, edited by W. Blackstone, 43–68. Athens, GA: University of Georgia Press.

Feinberg, J. 1984. *The Moral Limits of the Criminal Law*. Vol. 1, *Harm to Others*. Oxford: Oxford University Press.

Feldman, F. 1991. "Some Puzzles about the Evil of Death." *Philosophical Review* 100, no. 2: 205–27.

Feldman, F. 1992. *Confrontations with the Reaper: A Philosophical Study of the Nature and Value of Death*. New York: Oxford University Press.

Feldman, F. 2000. "Basic Intrinsic Value." *Philosophical Studies* 99: 319–46.

Fischer, J., ed. 1993. *The Metaphysics of Death*. Stanford, CA: Stanford University Press.

Flegenheimer, M. 2013. "As Rats Persist, Transit Agency Hopes to Curb Their Births." *New York Times*, March 11.

Foer, Jonathan Safran. 2009. *Eating Animals*. New York: Little, Brown.

Francione, G. 2008 . *Animals as Persons*. New York: Columbia University Press.

Francione, G., and R. Garner. 2010. *The Animal Rights Debate*. New York: Columbia University Press.

Franco, N. H., M. Magalhãe s-Sant'Ana, and I. A. S. Olsson. 2014. "Welfare and Quantity of Life." In *Dilemmas in Animal Welfare*, edited by M. C. Appleby, D. M. Weary, and P. Sandøe, 46–66. Wallingford, UK: CABI International.

Frank, J., and P. Frank. 2007. "Analysis of Programs to Reduce Overpopulation of Companion Animals: Do Adoption and Low Cost Spay/Neuter Programs Merely Cause Substitution of Sources?" *Ecological Economics* 62: 740–46.

Frey, R. 1977. "Animal Rights." *Analysis* 37: 186–89.

Frey, R. 1983. *Rights, Killing and Suffering*. Oxford: Clarendon.

Frick, K., P. Gloor, and D. Gürtler. 2013. "Global Thought Leaders 2013." Zurich: Gottlieb Duttweiler Institute. http://www.gdi.ch/de/Think-Tank/Global-Thought-Leaders-2013.

Garner, R. 2013. *A Theory of Justice for Animals: Animal Rights in a Non-ideal World*. New York: Oxford University Press.

Gerlai, R. 2012. "Using Zebrafish to Unravel the Genetics of Complex Brain Disorders." *Behavioral Neurogenetics* 12: 3–24.

Gewirth, A. 1978. *Reason and Morality*. Chicago: University of Chicago Press.

Gewirth, A. 1982. "There Are Absolute Rights." *Philosophical Quarterly* 32, no. 129: 348–53.

Gjerris, M., M. E. J. Nielsen, and P. Sandøe. 2013. *The Good, the Right and the Fair: An Introduction to Ethics*. Texts in Philosophy, no. 22. London: College Publications.

Glock, H. 2009. "Can Animals Act for Reasons?" *Inquiry* 52: 232–54.

Glover, J. 2006. *Choosing Children: Genes, Disability and Design*. Oxford: Oxford University Press.

Griffin, D. 1976. *The Question of Animal Awareness*. New York: Rockefeller University Press.

Griffin, J. 1986. *Well-Being: Its Meaning, Measurement and Moral Importance*. Oxford: Clarendon.

Griffin, J. 2008. *On Human Rights*. Oxford: Oxford University Press.

Habermas, J. 1990. *Moral Consciousness and Communicative Action*. Cambridge, MA: MIT Press.

Habermas, J. 1993. *Justification and Application: Remarks on Discourse Ethics*. Cambridge, MA: MIT Press.

Habermas, J. 1996, *Between Facts and Norms: Contributions to a Discourse Theory of Law and Democracy*. Cambridge: Polity Press.

Hare, C. 2007. "Voices from Another World: Must We Respect the Interests of People Who Do Not, and Will Never Exist?" *Ethics* 117, no. 3: 498–523.

Hare, R. 1993. "Possible People." In *Essays on Bioethics*. Oxford: Clarendon.

Hare, R. 1993. "Why I Am Only a Demi-Vegetarian." In *Singer and His Critics*, edited by Dale Jamieson. Oxford: Blackwell pp. 233–246.

Harish. 2012. "How Many Animals Does a Vegetarian Save." Counting Animals. http://www.countinganimals.com/how-many-animals-does-a-vegetarian-save/.

Harman, E. 2011. "The Moral Significance of Animal Pain and Animal Death." In *The Oxford Handbook of Animal Ethics*, edited by T. L. Beauchamp and R. G. Frey, 726–37. Oxford: Oxford University Press.

Harman, G. 1967. "Toward a Theory of Intrinsic Value." *Journal of Philosophy* 64: 792–804.

Harrison, R. 1964. *Animal Machines*. London: Vincent Stuart.

Haydon, D. T., R. K. Rowland, and R. P. Kitching. 2004. "The UK Foot-and-Mouth Disease Outbreak—The Aftermath." *Nature Review Microbiology* 2: 675–81.

Heathwood, C. 2005. "The Problem of Defective Desires." *Australasian Journal of Philosophy* 83: 487–504.

Herman, B. 1993. *The Practice of Moral Judgment*. Cambridge, MA: Harvard University Press.

Herstein, O. 2013. "Why 'Nonexistent People' Do Not Have Zero Wellbeing but No Wellbeing At All". *Journal of Applied Philosophy* 30 (2): 136–45.

Heyd, D. 1992. *Genethics: Moral Issues in the Creation of People*. Berkeley: University of California Press.

Hobbes, T., 1994. *Leviathan*. Edited by Curley Edwin. Indianapolis, IN: Hackett.

Holtug, N. 2001. "On the Value of Coming into Existence." *Journal of Ethics* 5, no. 4: 361–84.

Holtug, N. 2004. "Person-Affecting Moralities." In *The Repugnant Conclusion: Essays on Population Ethics*, edited by Jesper Ryberg and Torbjörn. Tännsjö. Dordrecht, The Netherlands: Kluwer Academic , pp. 129–162.

Holtug, N. 2007. "Animals: Equality for Animals." In *New Waves in Applied Ethics*, edited by Jesper Ryberg, Thomas Sóbirk Petersen, and Clark Wolf, 1–24. Basingstoke, UK: Palgrave MacMillan.

Holtug, N. 2010. *Persons, Interests, and Justice*. Oxford: Oxford University Press.

Holtug, N. 2011. "Killing and the Time-Relative Interest Account." *Journal of Ethics* 15, no. 3: 169–89.

HSUS (Humane Society of the United States). 2014. *Pets by the Numbers*. Washington, DC: HSUS. http://www.humanesociety.org/issues/pet_overpopulation/facts/pet_ownership_statistics.html.

Hume, D. 1975. "Enquiry Concerning the Principles of Morals." In *Enquiries Concerning Human Understanding and Concerning the Principles of Morals*, 3rd ed., edited by L. Selby-Bigge and P. Nidditch. Oxford: Clarendon.

Illies, C. 2003. *The Grounds of Ethical Judgement: New Transcendental Arguments in Moral Philosophy*. Oxford: Oxford University Press.

Jamieson, D. 1993. "Great Apes and the Human Resistance to Equality." In *The Great Ape Project*, edited by P. Cavalieri and P. Singer, 223–27. New York: St Martin's Griffin.

Johansson, J. 2010. "Being and Betterness." *Utilitas* 22, no. 3: 285–302.

Kain, M. 2010. "Duties Regarding Animals." In *Kant's Metaphysics of Morals: A Critical Guide*, edited by L. Denis, 210–33. Cambridge: Cambridge University Press.

Kaldewaij, F. 2008. "Animals and the Harm of Death." In *The Animal Ethics Reader*, edited by S. Armstrong and R. Botzler, 59–62. 2nd ed. London: Routledge.

Kant, I. 1900–. *Kants Gesammelte Schriften*. Edited by Royal Prussian (later German) Academy of Sciences. Berlin: George Reimer(later Walter de Gruyter).

Kant, I. 1987. *Critique of Judgment*. Translated by S. Pluhar Werner. Indianapolis, IN: Hackett.

Kant, I. 1991. "Conjectures on the Beginnings of Human History." In *Kant's Political Writings*. 2nd ed. Translated by H. Nisbet, edited by Hans Reiss. Cambridge: Cambridge University Press.

Kant, I. 1996. *The Metaphysics of Morals*. Translated and edited by M. Gregor. Cambridge Texts in the History of Philosophy. Cambridge: Cambridge University Press.

Kant, I. 1997a. *Critique of Practical Reason*. Translated and edited by M. Gregor. Cambridge Texts in the History of Philosophy. Cambridge: Cambridge University Press.

Kant, I. 1997b. *Lectures on Ethics*. Translated and edited by P. Heath and J. Schneewind. New York: Cambridge University Press.

Kant, I. 1998. *Groundwork of the Metaphysics of Morals*. Translated and edited by M. Gregor. Cambridge Texts in the History of Philosophy. Cambridge: Cambridge University Press.

Kant, I. 2000. *The Metaphysics of Morals*. Cambridge: Cambridge University Press.

Keeling, M. J., M. E. J. Woolhouse, R. M. May, G. Davies, and B. T. Grenfell. 2003. "Modelling Vaccination Strategies against Foot-and-Mouth Disease." *Nature* 421: 136–42.

Klocksiem, J. 2012. "A Defense of the Counterfactual Comparative Account of Harm." *American Philosophical Quarterly* 49, no. 4: 285–300.

Korsgaard, C. 1996a. "Kant's Formula of Humanity." In *Creating the Kingdom of Ends*, 106–32. New York: Cambridge University Press.

Korsgaard, C. 1996b. "Kant's Formula of Universal Law." In *Creating the Kingdom of Ends*, 77–105. New York: Cambridge University Press.

Korsgaard, C. 1996c. "The Reasons We Can Share." In *Creating the Kingdom of Ends*, 275–310. New York: Cambridge University Press.

Korsgaard, C. 1996d. *The Sources of Normativity*. Cambridge: Cambridge University Press.

Korsgaard, C., 2005. "Fellow Creatures: Kantian Ethics and Our Duties to Animals." In *The Tanner Lectures on Human Values*, Vol. 25, edited by G. Peterson, 77–110. Salt Lake City: University of Utah Press.

Korsgaard, C. 2007. "Autonomy and the Second Person Within: A Commentary on Stephen Darwall's *The Second-Person Standpoint*." *Ethics* 118, no. 1: 8–23.

Korsgaard, C. 2008. "Just Like All the Other Animals of the Earth." *Harvard Divinity Bulletin* 36, no. 3 (Autumn). http://www.people.fas.harvard.edu/~korsgaar/CMK.Just.Like.Other.Animals.pdf.

Korsgaard, C. 2009. *Self-Constitution: Agency, Identity, and Integrity*. Oxford: Oxford University Press.

Korsgaard, C. 2011a. "Interacting with Animals: A Kantian Account." In *The Oxford Handbook of Animal Ethics*, edited by T. Beauchamp and R. G. Frey, 91–118. Oxford: Oxford University Press.

Korsgaard, C. 2011b. "That Short but Imperious Word *Ought*: Human Nature and the Right." http://www.people.fas.harvard.edu/~korsgaar/Essays.htm.

Korsgaard, C. 2011c. "Human Beings and the Other Animals." Accessed October 1, 2014 http://www.people.fas.harvard.edu/~korsgaar/Essays.htm

Korsgaard, C. 2011d. "Valuing Our Humanity." http://www.people.fas.harvard.edu/~korsgaar/Essays.htm.

Korsgaard, C. forthcoming. "The Origin of The Good and Our Animal Nature." In *Problems of Goodness: New Essays on Metaethics*, edited by Bastian Reichardt. http://www.people.fas.harvard.edu/~korsgaar/CMK.MA1.pdf.

Laland, Keven. 2003. "Learning in Fishes: From Three-Second Memory to Culture." *Fish and Fisheries* 4 (3): 199–202.

Lassen, J., P. Sandøe, and B. Forkman. 2006. "Happy Pigs Are Dirty! Conflicting Perspectives on Animal Welfare." *Livestock Science* 103: 221–30.

Lau, H., T. Timbers, R. Mahmoud, and C. Rankin. 2013. "Genetic Dissection of Memory for Associative and Non-associative Learning in *Caenorhabditiselegans*." *Genes, Brain and Behavior* 12, no. 2: 210–23.

Lazari-Radek, K., and P. Singer. 2014. *The Point of View of the Universe: Sidgwick and Contemporary Ethics*. Oxford: Oxford University Press.

Leake, J. 2005. "The Secret Life of Moody Cows." *The Sunday Times*, February 27. http://www.thesundaytimes.co.uk/sto/news/uk_.news/article100199.ece.

Leenstra, F. R., G. Munnichs, V. Beekman, E. van den Heuvel-Vromans, L. H. Aramyan, and H. Woelders. 2011. "Killing Day-Old Chicks? Public Opinion Regarding Potential Alternatives." *Animal Welfare* 20: 37–45.

Lenman, J., and Y. Shemmer, Y. 2012. "Introduction." In *Constructivism in Practical Philosophy*, edited by J. Lenman and Y. Shemmer. Oxford: Oxford University Press.

Linzey, A. 2009. *Why Animal Suffering Matters*. Oxford: Oxford University Press.

Locke, J. 1975. *Essay Concerning Human Understanding*. Edited by Peter Nidditch. Oxford: Oxford University Press.

Locke, J. 1980. *Second Treatise of Government*. Edited by C. B. Macpherson. Indianapolis: Hackett.

Louise, J. 2006. "Right Motive, Wrong Action: Direct Consequentialism and Evaluative Conflict." *Ethical Theory and Moral Practice* 9, no. 1: 65–85.

Lucretius 1965. *On Nature*. Translated by Russel Geer. Indianapolis: Bobbs-Merrill.

Luper, S. 2007. "Mortal Harm." *Philosophical Quarterly* 57, no. 227: 239–51.

Luper, S. 2009a. *Philosophy of Death*. Cambridge: Cambridge University Press.

Luper, S. 2009b. Review of *Well-Being and Death*, by Ben Bradley. *Notre Dame Philosophical Reviews*. http://ndpr.nd.edu/review.cfm?id=16606.

Luper, S. 2012. "Exhausting Life." *Journal of Ethics: An International Philosophical Review* 16, no. 3: 1–21.

Luper, S. 2014a. "Life's Meaning," in Steven Luper, ed., *Cambridge Companion to Life and Death*, Cambridge University Press, 198–212.

Luper, S. 2014b. "Persimals." In *Spindel Supplement: The Lives of Human Animals*, edited by Stephen Blatti, special issue, *Southern Journal of Philosophy*, 52, Suppl. S1 (September): 140–62.

Marquis, D. 1989. "Why Abortion Is Immoral." *Journal of Philosophy* 86: 183–203.

Mason, G., and K. Littin. 2003. "The Humaneness of Rodent Pest Control." *Animal Welfare* 12: 1–37.

Matheny, G., and K. Chan. 2005. "Human Diets and Animal Welfare: The Illogic of the Larder." *Journal of Agricultural and Environmental Ethics* 18, no. 6: 579–94.

McCloskey, H. 1979. "Moral Rights and Animals." *Inquiry* 22: 23–54.

McDaniel, K., and B. Bradley. 2008. "Desires." *Mind* 117: 267–302.

McMahan, J. 1988. "Death and the Value of Life." *Ethics* 99, no. 1: 32–61. Reprinted in Fischer, J., ed. 1993. *The Metaphysics of Death*. Stanford: Stanford University Press.

McMahan, J. 1998. "Preferences, Death and the Ethics of Killing." In *Preferences*, edited by C. Fehige and U. Wessels, 471–502. Berlin: Walter de Gruyter.

McMahan, J. 2002. *The Ethics of Killing*. Oxford: Oxford University Press.

McMahan, J. 2013. "Causing People to Exist and Saving People's Lives." *Journal of Ethics* 17: 5–35.

Meacham, C. 2012. "Person-Affecting Views and Saturating Counterpart Relations." *Philosophical Studies* 158: 257–87.

Mench, J. A. 1998. "Thirty Years after Brambell: Whither Animal Welfare Science?" *Journal of Applied Animal Welfare Science* 1: 91–102.

Midgley, M. 1983. *Animals and Why They Matter*. Athens, GA: University of Georgia Press.

Miele, M., I. Veissier, A. Evans, and R. Botreau. 2011. "Animal Welfare: Establishing a Dialogue between Science and Society." *Animal Welfare* 20: 103–17.

Miller, D. 2008. "Political Philosophy for Earthlings." In *Political Theory: Methods and Approaches*, edited by D. Leopold and M. Stears, 29–48. Oxford: Oxford University Press.

Mohr, Noam. 2012. "Average and Total Numbers of Land Animals Who Died to Feed Americans in 2011." Machipango, VA: United Poultry Concerns. http://www.upc-online.org/slaughter/2011americans.pdf.

Nagel, T. 1978. *The Possibility of Altruism*. Princeton, NJ: Princeton University Press.

Nagel, T. 1993. "Death." In *The Metaphysics of Death*, edited by John Martin Fischer, 59–69. Stanford, Calif.: Stanford University Press.

Narveson, J. 1967. "Utilitarianism and New Generations." *Mind* 76, no. 301: 62–72.

Ng, Y. 1989. "What Should We Do About Future Generations? Impossibility of Parfit's Theory X" In *Economics and Philosophy* 5: 235–253.

Norcross, A. 1999. "Intransitivity and the Person-Affecting Principle." *Philosophy and Phenomenological Research* 59, no. 3: 767–76.

Norcross, A. 2012. "The Significance of Death for Animals." In *The Oxford Handbook of Philosophy of Death*, edited by Ben Bradley, Fred Feldman, and Jens Johansson. Oxford: Oxford University Press, pp. 465–474.

Nozick, R. 1974. *Anarchy, State and Utopia*. New York: Basic Books.

Nozick, R. 1981. *Philosophical Explanations*. Cambridge, MA: Harvard University Press.

Nussbaum, M. 2006. *Frontiers of Justice: Disability, Nationality, Species Membership*. Cambridge, MA: Belknap Press of Harvard University Press.

Olds, J. 1955. "'Reward' from Brain Stimulation in the Rat." *Science* 122: 878.

Olds, J. 1958. "Self-Stimulation of the Brain." *Science* 127: 315–24.

Olson, E. 1997. *The Human Animal*. Oxford: Oxford University Press.

Olson, E. 2002. "What Does Functionalism Tell Us about Personal Identity?" *Noûs* 36, no. 4: 682–98.

Olson, E. 2007. *What Are We? A Study in Personal Ontology*. Oxford: Oxford University Press.

Olsson, I., N. Franco, D. Weary, and P. Sandøe. 2012. "The 3Rs Principle—Mind the Ethical Gap." *ALTEX Proceedings* 1: 333–36.

O'Neill, O. 1989. *Constructions of Reason: Explorations of Kant's Practical Philosophy*. Cambridge: Cambridge University Press.

O'Neill, O. 2002. "Constructivism in Rawls and Kant." In *The Cambridge Companion to Rawls*, edited by S. Freeman, 347–67. Cambridge: Cambridge University Press.

Ord, T. 2008. "How to Be a Consequentialist about Everything?" Paper presented at the Tenth Conference of the International Society for Utilitarian Studies at University of California.

Parfit, D. 1984. *Reasons and Persons*. Oxford: Oxford University Press.

Parfit, D. 2012. "We Are Not Human Beings." *Philosophy* 87: 5–28.

Perry, C., and A. Barron. 2013. "Neural Mechanisms of Reward in Insects." *Annual Review of Entomology* 58: 543–62.

Persson, I. 1997. "Person-Affecting Principles and Beyond." In *Contingent Future Persons*, edited by N. Fotion and J. Heller. Dordrecht, The Netherlands: Kluwer Academic.

Persson, I. 2004. "The Root of the Repugnant Conclusion and its Rebuttal." In *The Repugnant Conclusion: Essays on Population Ethics*, edited by J. Ryberg and Torbjörn Tännsjo. Dordrecht, The Netherlands: Kluwer Academic.

PETA (People for the Ethical Treatment of Animals). "The Hidden Lives of Cows." http://www.peta.org/issues/animals-used-for-food/factory-farming/cows/hidden-lives-cows/.

Pollan, M. 2006. *The Omnivore's Dilemma. A Natural History of Four Meals*. New York: Penguin.

Quinn, W. 1984. "Abortion: Identity and Loss." *Philosophy and Public Affairs* 13.1: 24–54.

Rachels, J. 1990. *Created from Animals: The Moral Implications of Darwinism*. Oxford: Oxford University Press.

Rachels, S. 1998. "Counterexamples to the Transitivity of Better Than." *Australasian Journal of Philosophy* 76: 71–83.

Rachels, S. 2011. "Vegetarianism." In *The Oxford Handbook of Animal Ethics*, edited by Tom Beauchamp and R. G. Frey. Oxford: Oxford University Press, pp. 877–905.

Rawls, J. 1971. *A Theory of Justice*. Cambridge, MA: Harvard University Press.

Rawls, J. 2001. *The Law of Peoples*. Cambridge, MA: Harvard University Press.

Raz, J. 1988. *The Morality of Freedom*. Oxford: Clarendon.

Regan, T. 1976. "McCloskey on Why Animals Cannot Have Rights." *Philosophical Quarterly* 26: 251–57.

Regan, T. 2004. *The Case for Animal Rights*. Updated with a new preface. Berkeley: University of California Press.

Roberts, M. A. 2003. "Can It Ever Be Better Never to Have Existed at All? Person-Based Consequentialism and a New Repugnant Conclusion." *Journal of Applied Philosophy* 20: 159–85.

Roberts, M. A. 2009. "The Nonidentity Problem." In *The Stanford Encyclopedia of Philosophy*, edited by Edward N. Zalta. Stanford, CA: Stanford University. http://plato.stanford.edu/entries/nonidentity-problem/.

Rollin, B. E. 1995. *The Frankenstein Syndrome: Ethical and Social Issues in the Genetic Engineering of Animals*. Cambridge: Cambridge University Press.

Rollin, B. E. 1998. "On Telos and Genetic Engineering." In *Animal Biotechnology and Ethics*, edited by A. Holland and A. Johnson, 156–71. New York: Springer.

Rorty, A. 1983. "Fearing Death." *Philosophy* 58: 175–88.

Rosenbaum, S. 1993. "How to Be Dead and Not Care: A Defense of Epicurus." In *The Metaphysics of Death*, edited by J. Fischer, 117–34. Stanford, CA: Stanford University Press.

Rousseau, J. 1987. *On the Social Contract*. Edited by D. Cress. Indianapolis, IN: Hackett.

Rowlands, M. 2002. *Animals Like Us*. London: Verso.

Russell, W. M., and R. L. Burch. 1959. *The Principles of Humane Experimental Technique*. London: Methuen.

Sachs, B. 2011. "The Status of Moral Status." *Pacific Philosophical Quarterly* 91, no. 1: 87–104.

Salt, H. 1914. "Logic of the Larder." www.henrysalt.co.uk/bibliography/essays/logic-of-the-larder.

Sandøe, P. 2011. "Welfare." In *The Encyclopedia of Applied Animal Behaviour and Welfare*, edited by D. S. Mills, 642–43. Wallingford, UK: CABI Press.

Sandøe, P., and K. K. Jensen. 2011. "The Idea of Animal Welfare: Developments and Tensions." In *Veterinary and Animal Ethics: Proceedings of the First International Conference on Veterinary and Animal Ethics*, edited by C. M. Wathes, S. A. Corr, S. A. May, S. P. McCulloch, and M. C. Whiting, 19–31. West Sussex, UK: Blackwell.

Sapontzis, S. 1987. *Morals, Reason, and Animals*. Philadelphia: Temple University Press.

Saving 90. http://www.saving90.org/.

Scarlett, J. M., M. Salman, M. J. New, and P. Kass. 1999. "Reasons for Relinquishment of Pets in U.S. Shelters: Lifestyle Issues and Allergies." *Journal of Applied Animal Welfare Science* 2: 41–57.

Sen, A. 2009. *The Idea of Justice*. London: Allen Lane.

Sher, G. 1997. *Approximate Justice: Studies in Non-ideal Theory*. Lanham MD: Rowman & Littlefield.

Shiffrin, S. 1999. "Wrongful Life, Procreative Responsibility, and the Significance of Harm." *Legal Theory* 5, no. 2: 117–48.

Shoemaker, S. 1984. "Personal Identity: A Materialist's Account." In *Personal Identity*, edited by Sidney Shoemaker and Richard Swinburne. Oxford: Blackwell, pp. 67–132.

Sidgwick, H. 1907. *The Methods of Ethics*. 7th ed. London: Macmillan.

Simmons, J. 2010. "Ideal and Nonideal Theory." *Philosophy and Public Affairs* 38, no. 1: 5–36.

Singer, P. 1975. *Animal Liberation*. New York: Random House.

Singer, P. 1979. *Practical Ethics*. Cambridge: Cambridge University Press.

Singer, P. 1986. "All Animals are Equal." In *Applied Ethics*, edited by P. Singer, 215–28. Oxford: Oxford University Press.

Singer, P. 1993. *Practical Ethics*. 2nd ed. Cambridge: Cambridge University Press.

Singer, P. 1995. *Animal Liberation*. 2nd ed. London: Pimlico.

Singer, P. 2009. *Animal Liberation*. New York: Harper Perennial.

Singer, P. 2011. *Practical Ethics*. 3rd ed. Cambridge: Cambridge University Press.

Smith, J., and K. Boyd, eds. 1991. *Lives in the Balance: The Ethics of Using Animals in Biomedical Research*. New York: Oxford University Press.

Smith, M. T., A. M. Bennett, M. J. Grubman, and B. C. Bundy. 2014. "Foot-and-Mouth Disease: Technical and Political Challenges to Eradication." *Vaccine*, 32: 3902–908.

Snowdon, P. 1990. "Persons, Animals, and Ourselves." In *The Person and the Human Mind*, edited by Christopher Gill. Oxford: Oxford University Press, pp. 83–107.

Steinbock, B. 1978. "Speciesism and the Idea of Equality." *Philosophy* 53: 247–56.

Steiner, H. 2005. "Moral Rights." In *The Oxford Handbook of Ethical Theory*, edited by D. Copp, 459–80. Oxford: Oxford University Press.

Stemplowska, Z. 2008. "Worth the Paper It's Written On." *Journal of Political Ideologies* 13, no. 3: 228–33.

Stephen, L. 1896. *Social Rights and Duties*. London: Swan Sonnenschein, 1896. Available at http://www.gutenberg.org/files/28901/28901-h/28901-h.htm.

Street, S. 2008. "Constructivism about Reasons." In *Oxford Studies in Metaethics*. Vol. 3, edited by R. Shafer-Landau, 207–45. Oxford: Clarendon.

Street, S. 2010. "What Is Constructivism in Ethics and Metaethics?" *Philosophy Compass* 5, no. 5: 363–84.

Street, S. 2012. "Coming to Terms with Contingency: Humean Constructivism about Practical Reason." In *Constructivism in Practical Philosophy*, edited by J. Shemmer and Y. Lenman, 40–59. Oxford: Oxford University Press.

Sztybel, D. 2000. "Response to Everlyn Pluhar's 'Non-Obligatory Anthropocentrism.'" *Journal of Agricultural and Environmental Ethics* 13: 337–40.

Tasioulas, J. 2002. "Human Rights, Universality, and the Values of Personhood: Retracing Griffin's Steps." *European Journal of Philosophy* 10: 79–100

Temkin, L. S. 1987. "Intransitivity and the Mere Addition Paradox." *Philosophy and Public Affairs* 16: 138–87.

Temkin, L. S. 1993. *Inequality*. New York: Oxford University Press.

Temkin, L. S. 1999. "Intransitivity and the Person-Affecting Principle: A Response." *Philosophy and Phenomenological Research* 59, no. 3: 777–84.

Temkin, L. S. 2000. "Equality, Priority and the Levelling Down Objection." In *The Ideal of Equality*, edited by M. Clayton and A. Williams, 126–161. New York: Palgrave.

"Temple Grandin." In *Wikipedia*. http://en.wikipedia.org/wiki/Temple_Grandin. Accessed September 9, 2014.

Tildesley, M. J., N. J. Savill, D. J. Shaw, R. Deardon, S. P. Brooks, M. E. J. Woolhouse, and M. J. Keeling. 2006. "Optimal Reactive Vaccination Strategies for a Foot-and-Mouth Outbreak in the UK." *Nature* 440: 83–6.

Timmermann, J. 2006. "Value without Regress: Kant's Formula of Humanity Revisited." *European Journal of Philosophy* 14, no. 1: 69–93.

Vallentyne, P. 2006. "Of Mice and Men: Equality and Animals." In *Egalitarianism: New Essays on the Nature and Value of Equality*, edited by Nils Holtug and Kasper Lippert-Rasmussen. Oxford: Clarendon, pp. 211–238.

Varner, G. 2012. *Personhood, Ethics, and Animal Cognition*. Oxford: Oxford University Press.

Velleman, D. 1991. "Well-Being and Time." *Pacific Philosophical Quarterly* 72: 48–77.

Velleman, D. 1993. "Well-Being and Time." In *The Metaphysics of Death*, edited by J. Fischer, 329–57. Stanford, CA: Stanford University Press. First published 1991 in *Pacific Philosophical Quarterly*.

Velleman, D. 2009. *How We Get Along*. Cambridge: Cambridge University Press.

Višak, T. 2013. *Killing Happy Animals: Explorations in Utilitarianism*. London: Palgrave Macmillan.

Warren, J. 2004. *Facing Death: Epicurus and his Critics*. Oxford: Clarendon.

Watts, J. 2009. "Chinese Try to Curb 'Plague of Desert Rats' in Tibet with Contraceptives." *The Guardian*, March 25. http://www.theguardian.com/environment/2009/mar/25/china-gerbils-deserts.

Williams, B. 1993. "The Makropulos Case: Reflections on the Tedium of Immortality." In *The Metaphysics of Death*, edited by J. Fischer, 73–92. Stanford, CA: Stanford University Press. First published 1973.

Wolf, C. 2009. "Do Future Persons Presently Have Alternate Possible Identities?" In *Harming Future Persons: Ethics, Genethics and the Nonidentity Problem*, edited by Melinda Roberts and David Wasserman. Dordrecht, The Netherlands: Springer, pp. 93–114.

Wood, A. 1999. "The Formula of Humanity as End in Itself." In *Kant's Ethical Thought*, 111–55. Cambridge: Cambridge University Press.

Yeates, J. W. 2010. "Death Is a Welfare Issue." *Journal of Agricultural and Environmental Ethics* 23: 229–41.

Zhang, Z. 2000. "Mathematical Models of Wildlife Management by Contraception." *Ecological Modelling* 132: 105–13.

Ypi, L. 2010. "On the Confusion between Ideal and Non-ideal Theory in Recent Debates on Global Justice." *Political Studies* 58 (3): 536–55.

INDEX